T0296369

The Heart of the Brain

The Heart of the Brain

The Hypothalamus and Its Hormones

Gareth Leng

The MIT Press
Cambridge, Massachusetts
London, England

This book was set in Stone Serif by Westchester Publishing Services.

Library of Congress Cataloging-in-Publication Data

Names: Leng, G. (Gareth), author.
Title: The heart of the brain : the hypothalamus and its hormones / Gareth Leng.
Description: Cambridge, MA : The MIT Press, [2018] | Includes bibliographical references and index.
Identifiers: LCCN 2017047175 | ISBN 9780262038058 (hardcover : alk. paper)
ISBN 9780262551939 (paperback)
Subjects: LCSH: Hypothalamus--Popular works. | Hypothalamic hormones--Popular works.
Classification: LCC QP383.7 .L46 2018 | DDC 573.4/59--dc23 LC record available at https://lccn.loc.gov/2017047175

For Arnaud, Trystan, Rhodri, and Owain, and for Nancy

Contents

Preface

Clarity of thought distinguishes the best of scientists, and clarity of expression is particularly important in science, where fast and efficient communication underpins collective progress. Yet it is still an apparently widespread misconception that, for a scientific paper to be good, it must be dull, or obscure, or both. No referee or editor has ever advised me that a paper was unsuitable because it was too clear, too fluent, or too elegantly written.

This passage is from *Style Notes*, published as guidance to authors of papers submitted to The *Journal of Neuroendocrinology*, the leading specialist academic journal in my field. I wrote these in 1996 while I was editor in chief, and they have haunted me ever since as I have striven, always imperfectly, to live up to them.

I have written several hundred articles in the scientific literature, and every hour spent writing hides many others spent revising, and many, many more spent reading. Every thought I have sought to express, every discovery related, including those I would claim as my own, owes debts to many others. Mindful of the response of a reviewer to one scientific monograph, that the author had used references like a dog uses a lamppost—to mark rather than to illuminate, I have kept the number of references down: most chapters encapsulate arguments detailed more fully in review articles that I have cited, where I have more fully (but still incompletely) acknowledged those debts.

There can be no comprehensive acknowledgment of those debts. The world of Science is a commonwealth, within which, by and large, knowledge, expertise, and ideas flow freely. Tracing the "true" origin of any idea is like asking of a drop of water in a river from which cloud it fell, and about as useful.

In reading again for this book, I paid close attention to the way in which colleagues whom I respect for their science write. Good style, when reason flows seamlessly, is often invisible to the reader. It is not rare in academic science writing. But because such writing is intended for an expert audience, its virtues may not be apparent to all. The poet embraces ambiguities, builds layers upon layers of meaning with them, and imposes a further level of meaning by choosing words whose sounds and rhythms confer an emotional salience. The scientist-writer must avoid ambiguities, not exploit them, and this imperative often leads to the choice of technical terms rather than words in general use. The intended audience is likely to be familiar with these, but, being of many nationalities, might be unfamiliar with other words or phrases well known to those who speak English as a first language. The scientist-writer must therefore heed the advice of Samuel Johnson as related by James Boswell: "Read over your compositions, and where ever you meet with a passage which you think is particularly fine, strike it out."

I am indebted to many colleagues and friends around the world, whose passion for science has helped sustain my own: among these are many, including Françoise Moos, Susan Wray, Glenn Hatton, Valery Grinevich, John Morris, and Allan Herbison, whose work I have highlighted. My long and close friendships and collaborations with John Bicknell, Suzanne Dickson, Alison Douglas, Richard Dyball, Mike Ludwig, Duncan MacGregor, John Russell, Celine Caquineau, and Nancy Sabatier have been a constant support—with each of these I worked closely for ten or more years. Many others have worked with me for shorter times, including Colin Brown, Yoichi Ueta, Simon Luckman, and Louise Johnstone, whose work I also mention here. Others who have worked in my lab have come from many countries: France, Germany, Spain, Sweden, the Netherlands, Hungary, Japan, Australia, Thailand, India, Australia, Chile, and the United States. I have been lucky to be part of a commonwealth of ideas.

I particularly thank John Menzies, Nancy Sabatier, Mike Ludwig, Kevin O'Byrne, Anne Duittoz, Duncan MacGregor, and my sons Rhodri and Trystan for reading and commenting insightfully on sections of this book. The errors are of course all mine.

1 Prelude

Technical skill is mastery of complexity while creativity is mastery of simplicity.
—Chris Zeeman (1926–2016)[1]

I, like you, am a human being, a *Homo sapiens*. We share the same genes, with a few minor differences, but those genes did not solely determine who we are now, nor did the differences solely determine the ways in which we differ. The environment into which we are born and develop, our early life experiences and our interactions with close kin, make a big and lasting difference. I was born in Britain, with English my first language. I was born into an academic family: my father was a sociologist, my brother a computer scientist, my sister a teacher; our children include three doctors, a teacher, a historian, a philosopher, a mathematician, a psychologist, and a political scientist.

It might have been different. My father was the Welsh-speaking son of a miner, and the only one of his extended family to go to university. After six years as a soldier during World War II, he taught French at a secondary school in Manchester before returning to Wales to join the Department of Education at Bangor University, where he introduced sociology, philosophy, and psychology into the curriculum. My mother was one of eight surviving children. When her father died young, her mother eked out a living by taking in the washing from her neighbors; she was supported by her only son, my uncle Will, who left school at 14 to work in the coal mines of Aberdare. My mother left school at 16 to work in a library, a job she had to give up when she married, and her life thereafter was that of a homemaker.

By knowing these things that are things not about me but about my close kin you might make a host of predictions about me, about my politics, social attitudes, lifestyle and personality, even about what sports I follow and the books I read. Most would probably be accurate. Yet still my life might easily have been different. I studied mathematics at the University of Warwick, but lost the passion for it that I began with. As the end of my time at University beckoned, I applied for three jobs, as a teacher, as a technical editor, and as a systems analyst trainee, and luckily was rejected for all three. My careers advisor suggested that I try to write for *Rolling Stone*, which shows how lost I was and how helpless he was. My encounter with one person changed things: Chris Zeeman, the head of the Maths Department, inspired me to think that mathematics could find simple explanations for complex behaviors, and he planted in me the thought that biology might be fun. I went to Birmingham to learn some neuroscience, and three years later I was an electrophysiologist with a PhD in auditory physiology. Mathematics never quite left me, though I left it: its traces are in how I think about problems and in what I find interesting, and in later years I came back to mathematics in my work with computational modeling.

After my PhD, Barry Cross, the director of what was then the Agricultural Research Council's Institute of Animal Physiology and is now the Babraham Institute, recruited me as a project leader in his research group on neuroendocrinology. It was brave to appoint me: I didn't know what a hormone was, yet he gave me a lab, tenure, and freedom and trusted that I'd find out.

Barry Cross had been a student of Geoffrey Harris, the "founder" of neuroendocrinology who had shown how the hypothalamus controls the pituitary gland. Barry had been a pioneer in our understanding of oxytocin and he went on to be a pioneer in using electrophysiology to study the hypothalamus. When I arrived at Babraham, he told me, "Here's your lab. Let me know when you've found something interesting." Today, the path of scientists to independence is long and tenuous. They go from one short-term position to the next; they are expected to move to gain experience in different labs and to show that they can be productive in different environments and can win independent funding from a bitterly competitive system before even being considered for tenure in a research-strong institution. This system aspires to foster risk and innovation, but it rewards safety and conformity. In our conservative new world it is presumed that innovation is born from insecurity. This is false: risks are taken by those who can afford to fail, and in our conservative new world no one can afford to fail.

The nature of research has changed too, as technological expectations often require that scientists work in large groups, combining complementary expertise and collectively exploiting expensive equipment. This has meant the rise of the scientist as a technologist and as a manager of a research program, and the decline of the scientist as an independent thinker. This is not to say that ideas are less important than they were or that thinkers are less valued. Ideas should be the currency of the young and free, not the prerogative of the old and set.

I was lucky that, in my formative period, I had no boss and was nobody's boss. The collaborations that I entered into, first with John Bicknell, Richard Dyball, Richard Dyer, and John Russell, later with Mike Ludwig, Alison Douglas, and many others, were free collaborations among equals, and debates, disagreement, and dissent were our constant companions. Authority is the worst enemy of science; arguments by authority are no arguments at all. When my bluster and their cool reason were done, our arguments were settled by experiments that often went late into the night.

A scientist is a professional unbeliever, paid to be curious. There's nothing special about the scientific method: it's just practiced honesty. Perhaps scientists know more than most about the ways in which the world and our senses can fool us, but the world and our senses find new ways every day. Unless you are interested in the possibility of being wrong you will not be interested for long in science. Debates, disagreement, and dissent were for me the fuel for a passion that could endure. Science is about not certainty but doubt.

Authority is an insidious enemy of science; arguments by authority deserve no respect. I might have some authority regarding facts I have found myself, but these are small and insignificant in the scheme of things; the scientist gathers facts only as a means to an end, to spin them into a tale by wit and reason and imagination. I might use devices like statistics or mathematics to support my reason, but these things are just good sense made rigorous. Stripped of the jargon that we use among ourselves, our theories and explanations are usually things that any thinking person can follow. Our wit is supported by technical expertise, but this is not fundamentally different from the craft of the carpenter or gardener. I am no carpenter or gardener, but I know what a shelf is good for though the ones I make fall down, and what tomatoes are good for though my plants all wither. You don't have to know how the equipment in my lab works to understand what it does: it lets me measure the electrical signals made by nerve cells and the chemicals they release. I might be proud of my craft, as the cook and the lighthouse builder are proud of theirs, but I don't have to make it mystical.

So if facts once gathered are things that any may look on and judge, and if my theories are just common sense made careful, what I say should be understandable by any thinking person willing to listen. But being understandable is not the same as being right. All that we think we know might be wrong.

We all have prejudices, biases, and preconceptions. Scientists must put these aside as best they can, or at least be aware of them. We must be willing to go back to first principles if pressed. And if we find ourselves blustering and skipping, appealing to common knowledge, accepted opinion, received wisdom, usual interpretation, consensus, conventional understanding—then we will have learned something: either we have not thought things through as carefully as we should, or else the things we have taken for granted are wrong.

This book aims to do many things, probably too many. I wanted to celebrate the hypothalamus, the "lizard brain" as it is disparagingly called by some who think that neuroscience is just about consciousness and intelligence. The hypothalamus does mundane things, but it does them well, and how it solves difficult challenges can tell us much about what neurons and networks are capable of. But the hypothalamus has a greater influence on our behavior than we, who like to think, like to think. How much of what we do is really governed by reason? How often are the reasons that we give merely self-serving narratives, justifying behaviors that are governed by things of which we are unaware or only dimly aware, or that we prefer not to acknowledge?

I also wanted to display the imaginative part of science: how ideas begin, how theories arise from observations, exploration, and discovery, and how they are tested. Electrophysiologists like myself were caught up in the excitement of experiments. We would literally *listen* to our neurons: their electrical activity was broadcast on loudspeakers as we recorded it, and it often felt as if we were engaged in a conversation. We would ask questions by the experiments that we performed, and the neurons would answer. The conversations were often urgent; every neuron is different in some ways from every other, and each seems to have its own tale to tell. Often, we might have just the one chance to ask the important question of a particular neuron before its signal vanished into the ether, and finding the right question and the right way to ask it kept us on edge. Jon Wakerley, a colleague in my first years at Babraham, once confessed to me that he disliked holidays: on the first day he'd find himself looking at the tense and anxious faces of others scurrying to work and feel *envy*. My friend and collaborator Richard Dyball would laugh a lot but never told jokes that I remember—except one. A lawyer, a vicar, and an electrophysiologist were discussing the merits of having a mistress. The lawyer discounted it as expensive and the vicar as time-consuming, but the electrophysiologist was enthusiastic. "When I'm not at home," he said, "my wife assumes I'm with my mistress, and when I'm not with my mistress, *she* assumes I'm with my wife." And Richard blinked behind his large glasses, "and all the time I can be in the lab."

Many others didn't *get* electrophysiology; the frustrations and uncertain returns put them off. Others were immersed in the calm beauty of microscopic images, or in the infinite eccentricities of behavior, or in unfolding the patterns of development, or in the languages of genes and molecules. There are so many ways of being a scientist, and they call for so many different skills and personalities. Yet all ask for some inner spark, a spark that fires our imagination and drives us on.

Scientists in universities do not expect to become rich. Most of our students will, a few years on, earn more than us. We have extraordinary freedom, but it is a freedom given only to those who will not abuse it; though unshackled, we labor without rest. We devise and conduct experiments and analyze the results, but also guide, teach, and assess undergraduate students—and devise new courses for them. We train postgraduate students, and mentor young researchers. We write grant applications (most of which will go unfunded) and review the applications of others. We write papers for publication and revise or discard them in the light of the (usually just) criticisms of our peers, and review those of others. We spread our understanding at scientific conferences, to scientists in industry who might translate our research for human benefit, and in public lectures, science festivals, and blogs. We must keep up with a scientific literature that balloons daily, and technology that develops ever faster. We edit journals, organize conferences and workshops, and act as advisors to institutions large and small. And we help manage our universities, in a myriad of administrative tasks that are essential for balancing the books and keeping our students and colleagues safe and sane.

Yet we are rich. There is an absurd notion about that education and health are things that we spend our national wealth on. What an extraordinary idea this is that money has any value in itself; education and health are not things we spend our wealth on, they *are* our wealth. To be a scientist, to be a creator of knowledge and understanding, is to be a creator of wealth in the most important and enduring sense.

Keats drank confusion to Newton for analyzing the rainbow,[2] and in his epic poem *Lamia*,[3] he fretted that science would drain the world of wonder:

Philosophy will clip an Angel's wings,
Conquer all mysteries by rule and line,
Empty the haunted air, and gnomed mine—
Unweave a rainbow, ...

Are we weavers or unweavers of rainbows? Wishing confusion on science is like wishing black on coal; the harder we look at anything the more confusion we find, and like poets we rely on imagination and metaphor, in our case to sieve some understanding from the slurry of facts that we dredge. But our metaphors need to be sound, and Keats's was weak: nobody who looks at a rainbow could ever conceive that it was woven, in any sense. Perhaps poets don't need to use their eyes,[4] but scientists must.

Shakespeare, as always, puts it best. His sonnet "My mistress' eyes are nothing like the sun" ends this way:[5]

My mistress, when she walks, treads on the ground:
And yet by heaven, I think my love as rare,
As any she belied with false compare.

2 The European Brain

"Owl," said Rabbit shortly, "you and I have brains. The others have fluff. If there is any thinking to be done in this Forest—and when I say thinking I mean thinking—you and I must do it."

—A. A. Milne (1882–1956), *The House at Pooh Corner*

Rabbit, the intellectual of the Forest, holds that the brain is for thinking. There is rather little evidence for this, though it may be true. The brain is, however, important for other things for which there is strong evidence, although we tend to neglect them as less important, or less mysterious, or less *human* than thinking. This scale of values is naturally to be endorsed by intellectuals, but I'm not sure that an inevitable logic underlies it.

When I was a child, one of my small pleasures was to eat fish and chips soaked in vinegar and salt. The fish and chips were wrapped in newspaper. My favorite was the *News of the World,* a newspaper that fed on scandal until it closed in 2011 in the wake of a scandal of its own, its journalists embroiled in allegations of greed, bribery, corruption, and sleaze.

The paper's motto was "All human life is there." The *News of the World* and its sister paper the *Sun* told the story of human life through stories of the extraordinary lives of ordinary people and the ordinary lives of extraordinary people. "The world's tallest man lives in Neasden" was a headline I remember, and I remember the woman who ate only crisps and the centenarian who smoked a pack of cigarettes every day. There were stories of fat people and thin people, large families and single mothers, gamblers, alcoholics, drug addicts and sex addicts. There was lots of sex: footballers romping in hotel rooms, three in a bed, secret love children and jealous wives, and lots of violence—the petty tantrums of the famous and nasty tantrums of the not-so-famous—road rage, the revenge of lovers scorned, the despoilation of gardens, the theft of garden gnomes and the kidnapping of pets. The greed of the rich, the sloth of the poor, the fecklessness of the young, the frauds and con-artists were celebrated alongside heroes and heroines who qualified for inclusion mainly by being unlikely.

What makes us human, by this accounting, is not our symphonies but our soap operas; not our ability to love, but our ability to seduce; not our poetry but our smooth tongues and spite; not our curiosity about the cosmos but our curiosity about our neighbors. Our absurdities and stupidities, hypocrisies and delusions, fallibilities and imperfections, not our reason, intelligence, and apprehension.

Despite claims that men think with their gonads, this is only approximately true. What *is* true is that our brains are very much concerned with sex. When we talk about The Brain we are sometimes thinking of the convoluted floppy bits that form the cerebral cortex. These are so much more extensive in humans than in other mammals that we are inclined to suspect that the explanation of our humanity and our intelligence lies in them. Perhaps they do explain our intelligence, but our humanity is entwined with our emotionality: our capacities for love and empathy, anxiety and disappointment; our experiences of fear, anger, and stress; our wayward passions and diverse appetites. How many of the decisions we make depend on our rationality? Behavioral economists know that even when we address problems eminently amenable to reason, such as where we should invest or how to spend our money, or when to gamble and when to hold, our decisions are subverted by a multitude of cognitive biases. Our human decisions—the big ones like whom to choose as a life partner, the sometimes big ones like when to fight and when to flee, and the smaller ones like what and when to eat and drink—these we *might* think about, but often only to wonder why we made the choices we did.

We might begin by asking where in the brain these decisions are made. Neuroscientists use maps of the brain, charting the anatomy, marking the paths that connect different regions, and they use these maps, with evidence from many sources, to attribute functions to particular parts. In the atlas of the human brain, the cerebral cortex is a bit like all the oceans of the world together, and about as featureless. By contrast, the hypothalamus is small, old, and gnarled, and what it does ranges from the mundane to the sublime. When we are dehydrated, it makes us thirsty and tells our kidneys not to waste water but to concentrate our urine. When we are ill, it raises our body temperature, generating a fever that kills off infections. When our blood sugar is low, it tells us to eat, and when our stomach is full, to stop. It determines the shape of our bodies, how tall we will grow, how fat we will be, and where our fat and muscle will grow. When we are frightened or anxious or stressed it determines whether we will set our teeth, stiffen our sinews, freeze, or fight—or flee. The rhythms of days and of seasons are beaten by its drums, and we grow and attain puberty under its tutelage. And, led by the hypothalamus, we preen and woo, and the sap rises in our loins; a man produces sperm, and in a woman the ovarian cycle turns; we court and mate and bond. A woman conceives, and her body changes to the needs of the child within, and she gives birth and produces milk, and she loves and nurtures her children.

A map of Europe is filled with small countries with aggressively different identities, their stupidities, absurdities, fallibilities and imperfections, grime and squalor jostling among pinnacles of art and beauty. The hypothalamus is the Europe of the brain, where different regions control gluttony, anger, sloth, obesity, drinking, addiction, stress, love and hate, and sex. All human life is there.

This book is about the hypothalamus and all the related parts of what we might call the *heart of the brain*. It tells stories of things that matter to all of us, but not *because* they matter. It tells these stories to build an argument.

The conventional view of the brain is as a sophisticated and massively complex computational machine. Each of 100 billion neurons communicates with 10,000 others up to 200 times per second, constantly changing those connections in the face of changing demands and experience. This vision sees neurons as essentially all reliable and all alike, differing mainly in whether the signals that they generate are inhibitory or excitatory. The connections between neurons create a web so complex that, even if we may never be able to understand exactly what is happening, its capacity for doing wonderful things needs no further explanation. It is a vision that seeks to understand the brain from its *connectome*—the precise organization of anatomical connections. The years have embellished this vision but the core has remained unchanged; in the brain, information is the cacophony of electrical signals in billions of axons triggering hundreds of billions of tiny chemical signals, from neuron to neuron in networks within networks within networks, and by some miracle explainable only by the incomprehensibility of this complexity, information is processed to yield coherent and sensible outcomes.

Neurons of the hypothalamus are *not* all alike; far from it. They comprise many subpopulations—tribes, if you like. Living as I do in Scotland, I'd rather think of clans.[1] Members of a clan are all different from each other, but are more like each other than like members of other clans. Each clan has its characteristic phenotype, dictated by the "tartan" of genes that it wears. Different genes make some neurons sensitive to glucose, temperature, or osmotic pressure, or to particular hormonal signals from the periphery. Others determine the signals that neurons of a clan generate. Different clans can be defined by the signals they use to communicate and the signals to which they can respond: different clans use different combinations of peptides as chemical signals along with "classical" neurotransmitters. These combinations come from more than a hundred known neuropeptides, and to these we should add yet more signaling molecules—prostaglandins, neurosteroids (steroids synthesized in the brain itself), endocannabinoids (endogenous, cannabis-like molecules), and gases like nitric oxide and carbon monoxide. Yet other genes determine where the clan is in the hypothalamus, and the shapes and connectivity of its members.

Some clans regulate the autonomic nervous system, which controls blood pressure, heart rate, digestion, respiration, urination, and sexual arousal. Some others are *neuroendocrine* neurons: these regulate the secretion of hormones from the pituitary gland—many of which control the secretion of *other* hormones, including those from the ovary and testes, the adrenal and thyroid glands—and also hormones from the liver, kidneys, gut, and heart. They control not only the functions of organs in our body, but also our behavior, by their actions on other parts of the brain. The hypothalamus of a male is not the same as that of a female—it is a sexually dimorphic structure. It is also plastic—its structure and functions are malleable, and alter according to physiological needs: after puberty, in pregnancy and lactation, in cold and hunger.

Behaviors important to who we are—love and hate, how much we eat and what we eat, how we respond to threat and to stress—are governed by the hypothalamus, and not by the map of how the neurons are connected, but by where the *receptors* for these peptide signals are found. Neurotransmitter signals are ephemeral and confined by anatomical connectivity, but the peptide signals that hypothalamic neurons generate are potent, wide-reaching, and long-lasting, and they affect not just neuronal signaling but also the genes that neurons express. Remarkably, different peptides when injected into the brain induce coherent, meaningful behaviors—some, for example, trigger eating, others induce a longing for salt or initiate maternal behavior or aggression or sleep.

As we began to recognize the complexity of the hypothalamus, the diversity of its systems and the scope of their effects, and the interactions between hormonal systems and behavior, the emerging "hypothalamo-centric" vision began to look different from the conventional view of the brain. This vision embraces the heterogeneity of neurons—the differences between neurons that do not merely or mainly reflect differences in how they are connected to other neurons. It also confronts the uncomfortable recognition that delivering a drug into the brain with no anatomical precision and no sophisticated pattern of delivery can elicit coherent behavioral outcomes. I find it hard to express just how shocking this is: it's as though you can take a sonnet of Shakespeare, cut out the words, and scatter them in the wind, only to find them gathered together faultlessly.

Individually, neurons are not perfect: they are erratic, messy, quarrelsome, and unreliable. But each clan organizes itself, and can do things that the individuals in it can't. The clans make light work of some apparently difficult tasks but struggle with other tasks that are apparently trivial: our intuitions are a poor guide. Clans talk to clans, and do so in many different ways with many types of signals on different spatial and temporal scales; they use not one language but many. Understanding those clans and their languages lets us see how patterns can emerge from apparent chaos, robustness from noise, decisions from quarrels, purposeful behavior from heterogeneity and confusion. Perhaps one day Europe will do as well.

3 The Classical Neuron

I met a traveller from an antique land
Who said: "Two vast and trunkless legs of stone
Stand in the desert. Near them on the sand,
Half sunk, a shattered visage lies, whose frown
And wrinkled lip and sneer of cold command
Tell that its sculptor well those passions read
Which yet survive, stamped on these lifeless things,
The hand that mocked them and the heart that fed."
—Percy Bysshe Shelley (1792–1822), "Ozymandias"

When anatomists began to study the fine structure of the brain, they saw that the neurons seemed to be connected by a dense mesh of fibers: thin fibers, *axons*, often traveled for long distances, while thicker fibers, *dendrites*, stayed close to the cell body and often branched extensively. By the end of the nineteenth century, Ramon y Cajal had shown that the neurons are not connected directly to each other but are physically separated.[1] Accordingly, for information to leave one neuron and be received by another, a message had to be released from the one, which was then recognized by the other. Although the fibers of different neurons did not form a continuous network, the axons often came very close to cell bodies and dendrites. By the early twentieth century it was recognized that these contacts included sites that we now know of as *synapses*.[2] Somehow, it seemed, messages crossed from neuron to neuron at these synapses.

The advent of electron microscopy made it possible to see things invisible to the light microscope, and it became clear that the synapse has many specialized features that, under close interrogation, eventually yielded their identities.

We now know that, at a synapse, the axon terminal is filled with many small vesicles that contain a chemical neurotransmitter, often the excitatory neurotransmitter *glutamate* or the inhibitory neurotransmitter *GABA* (gamma-amino butyric acid). These vesicles can be released by electrical signals which open pores in the terminal membrane to allow calcium ions to enter. This results in one or more vesicles fusing with the terminal membrane and emptying their content into the *synaptic cleft*, the narrow channel between the terminal and the dendrite (figure 3.1).

The other side of the synapse includes a complex structure, the *postsynaptic density*, which organizes the dendrite's response to chemical signals. This contains *receptors* for neurotransmitters, and it regulates their availability at the postsynaptic membrane. Receptors are proteins with a particular shape that allows certain other molecules to bind to them and thereby trigger some kind of signal. Some receptors are *ion channels*. When a neurotransmitter binds to one of these, the ion channel opens and a current flows into or out of the dendrite—these receptors convert chemical signals into electrical signals.

As in most cells, the cell body of a neuron contains the *nucleus* (which contains the cell's DNA), the *rough endoplasmic reticulum* (where peptides are assembled), and the *Golgi apparatus* (which packages peptides into vesicles). The fluid inside a neuron, as in all cells, is not the same as the fluid outside; it contains more potassium and less sodium, because the cell has *pumps* that import potassium while expelling sodium. Both sodium and potassium ions are positively charged, and, because the pumps expel more sodium than they import potassium, the neuron (like all cells) is electrically "polarized"—electrodes inside and outside it will measure a difference in voltage, with the inside of the cell negative relative to the outside; this difference is the cell's *membrane potential*. The normal "resting" membrane potential of a neuron is usually at about –70 mV relative to the outside of the cell. These differences between the inside and outside of a neuron are critical to understanding how neurotransmitters work. When glutamate binds to a receptor on the surface of a neuron, the shape of that receptor changes, and a pore opens to allow sodium ions through. Because the neuron is negatively polarized and its interior is low in sodium, a current carried by sodium ions will enter, raising ("depolarizing") the membrane potential by a small amount for a few milliseconds.

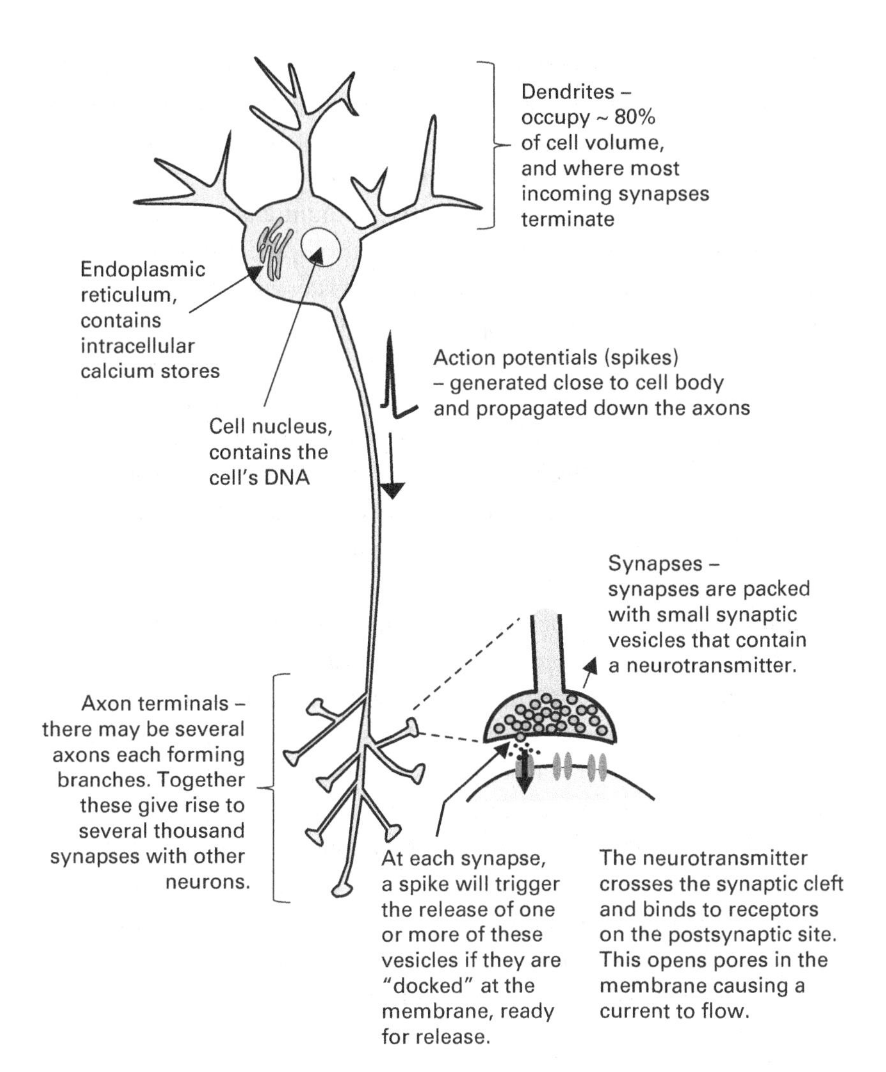

Figure 3.1
The classical neuron.

A typical neuron receives several thousand synaptic inputs from hundreds of neurons. The dendrites "integrate" these. If enough excitatory synapses are activated, the membrane potential will be depolarized beyond a critical point—the *spike threshold*. One site in the neuron, usually close to the cell body, contains abundant voltage-sensitive sodium channels, and when the neuron is depolarized enough, these start to open. As sodium enters, it depolarizes the neuron further, more channels open, and more sodium enters. The result is a voltage "spike"—an *action potential*. As this spike depolarizes the neuron, voltage-sensitive potassium channels open, potassium leaves the neuron, and the neuron repolarizes: in about a millisecond, the spike is over at its site of initiation. However, the spike travels down the axon as an electrical wave, on and on until it reaches the axon terminals. In 1963 Alan Hodgkin and Andrew Huxley were awarded a Nobel Prize for their role in understanding the mechanisms involved, and for producing a mathematical model that encapsulates that understanding.[3]

The information continuously received by a neuron is integrated across time (the effects of each input last a few milliseconds) and space (inputs arrive at different places on the dendrites and cell body). When the integrated signal exceeds the spike threshold, a spike is generated that travels down the axons: when it reaches the axon terminals, the spike will open calcium channels and the calcium entry will cause a neurotransmitter to be released. In this way, the chemical information received by a neuron is transformed into electrical signals, which are themselves resolved into a pattern of spikes that travel down the axon to generate *another* chemical signal that will be transmitted to other neurons. Thus, in a sense, spikes "encode" the information received by a neuron—but neurons talk to each other not by spikes themselves, but by chemical signals released by the spikes (figure 3.2).

For neuronal networks to become efficient at the tasks they execute, the synapses must "learn" from experience.[4] If a task requires one neuron to cause another to fire a spike, for that task to be fulfilled more efficiently experience must cause the synapse between the neurons to become stronger. This can be achieved either by the release of more neurotransmitter from the axon terminals or by an increase in the sensitivity of the dendrite. Both of these types of *synaptic plasticity* are common, as are many others. Synapses are not fixed forever: the neuronal circuits in our brain are continually being restructured throughout our lives. At every moment some synapses

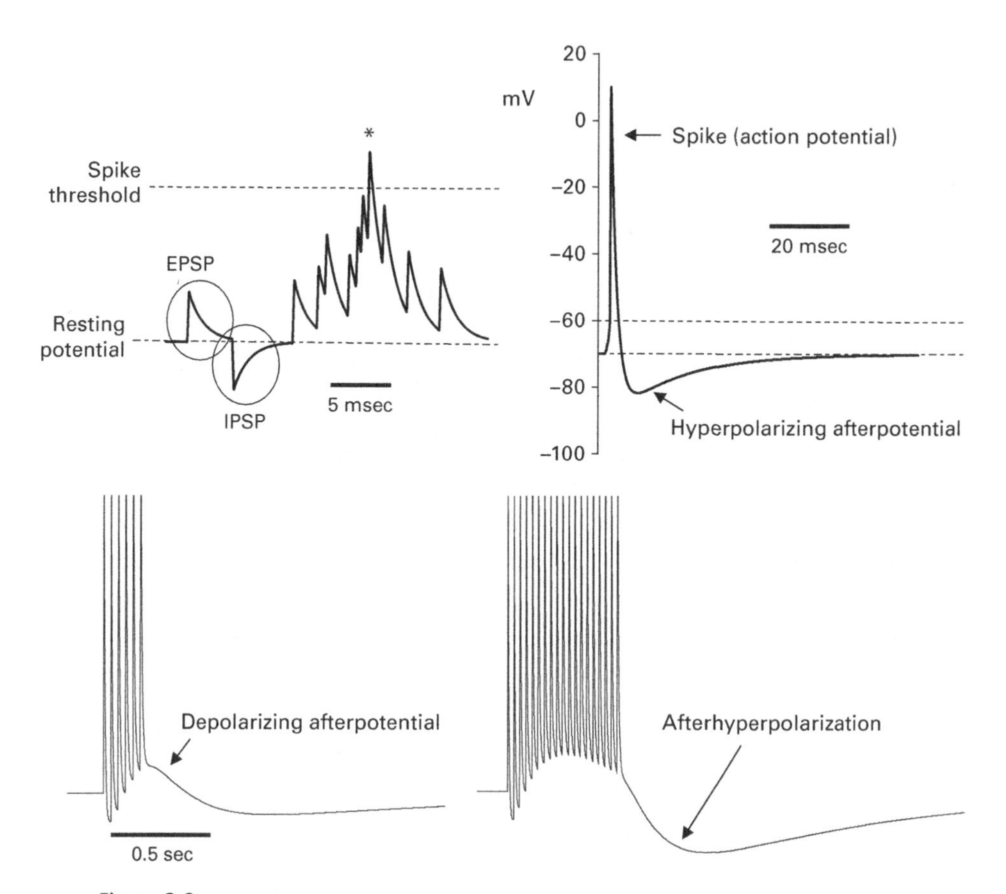

Figure 3.2

Generating spikes. Neurons are electrically *polarized*—at rest, the electrical potential inside a neuron is about 60–70 mV negative with respect to the outside. Neurotransmitters disturb this state by opening ion channels in the membrane, allowing brief currents to flow into or out of the neuron. The inhibitory transmitter GABA opens chloride channels—these cause negatively charged chloride ions to enter the cell (chloride is much more abundant in the extracellular fluid), and this current hyperpolarizes the neuron, causing an IPSP (an *inhibitory postsynaptic potential*). Conversely, the excitatory neurotransmitter glutamate causes sodium channels to open, and this causes the positively charged sodium ions to enter the neuron, depolarizing it and causing an EPSP (an *excitatory postsynaptic potential*). If a flurry of EPSPs causes a depolarization that exceeds the neuron's spike threshold, the neuron will fire a spike (an *action potential*). After a spike, neurons are typically inexcitable for a period because of a short *hyperpolarizing afterpotential*, and this limits how fast a neuron can fire. Trains of spikes can cause complex long-lasting changes in excitability—*depolarizing afterpotentials* or *hyperpolarizing afterpotentials* and combinations of these. These cause neurons to discharge in particular patterns of spikes. These properties vary considerably between different neuronal types.

are being pruned while others are proliferating: yesterday's brain was not quite today's.

Experience-dependent learning requires that some signal causes certain synapses to be strengthened. Many "retrograde" signals pass back from the dendrite to the axon terminal. One is nitric oxide; produced in many neurons when they are activated, this gas diffuses back into impinging axon terminals. Other retrograde signals include adenosine, prostaglandins, neurosteroids, and many neuropeptides. If a signal from one neuron is immediately followed by a spike in a neuron that it contacts, then a retrograde signal might strengthen the synapses between them, and this can make particular neuronal circuits more efficient. This is how, for neuronal networks, practice makes perfect. Conversely, if spike activity in two neurons is consistently *un*correlated then the synapses between them can become weakened or eliminated. Other "reward signals" from more distant neurons may signal that the flow of information through a network has been productive, and can reinforce all of the synapses in the chains that have been recently active. One of these signals is the neurotransmitter dopamine, released in the "reward circuits" of the brain, about which I will say more later.

This sketch of the brain, so thinly summarizing a century of endeavor, presents the brain as a computational structure whose power resides in its scale (the numbers of neurons and synapses), in its complexity (the connections that each neuron makes to other neurons), in the ability of neurons to compute rapidly (in milliseconds), and in the ability of neuronal networks to modify their connections in the light of experience.

By the end of the twentieth century an important refinement in our understanding of neurons had become accepted. The spike activity of neurons does *not* simply reflect the information received: it is also determined by their intrinsic properties, properties that vary from neuron to neuron.[5,6] Some neurons generate spikes regularly without any synaptic input: in these, synaptic activity does not determine the exact timing of spikes but modulates the average frequency of their discharge. Other neurons show prolonged activity-dependent changes in excitability. Some of these become more active as they are activated, so excitatory inputs trigger not single spikes but bursts of spikes; some show regular short bursts, others activity that waxes and wanes. Still others show long bursts separated by long silences. Sometimes the bursts begin with a peak of intensity, sometimes they end on a peak: the varieties seem endless.

Spike activity also depends on the mini-networks that neurons form with close neighbors. One common network motif comprises an excitatory neuron connected to an inhibitory interneuron that projects back to it; this can convert a continuous input into a transient output or into a sequence of bursts. Thus neurons, through their intrinsic properties and by the mini-networks that they form, transform the information they receive in diverse ways.

In 1959, Jerry Lettvin and colleagues published "What the Frog's Eye Tells the Frog's Brain" in the *Proceedings of the Institute of Radio Engineers.*[7] This paper became a classic both for the insight it provided and for its prose. Here is an extract from the introduction:

> The frog does not seem to see or, at any rate, is not concerned with the detail of stationary parts of the world around him. He will starve to death surrounded by food if it is not moving. His choice of food is determined only by size and movement. He will leap to capture any object the size of an insect or worm, providing it moves like one. He can be fooled easily not only by a bit of dangled meat but by any moving small object. His sex life is conducted by sound and touch. His choice of paths in escaping enemies does not seem to be governed by anything more devious than leaping to where it is darker. Since he is equally at home in water and on land, why should it matter where he lights after jumping or what particular direction he takes?

This paper showed that the outputs of the frog retina express the visual image in terms of (1) local sharp edges and contrast; (2) the curvature of edge of a dark object; (3) the movement of edges; and (4) the local dimmings produced by movement. In the retina, information processing involves discarding everything irrelevant to the two things that matter most: whether there is food to be had or danger to be avoided. This parsimony does not reflect anything simple or primitive about the frog's eye: the frog retina contains about a million receptor cells and about 4 million neurons, and frogs have been molded and tempered by the fires of fortune and natural selection for the same millennia that mammals have, and are better at being frogs than any mammal could be.

The processing involves resolving *edges* and *changes*. For neurons, to resolve an edge involves no more than comparing the activity of neurons to that of their immediate neighbors. To resolve changes is even more straightforward: neurons generally are good at responding to changes; retaining a steady signal is harder because when a neuron is excited by a constant stimulus, its activity generally declines. There can be many reasons for this: in the case of the photoreceptors in retinas, *adaptation* is a result of bleaching of the photopigment and is unavoidable.

Is our vision like that of the frog? Richard Dawkins, in *Unweaving the Rainbow* (1998), conceived that our retina, like that of the frog, only reports *changes* in what we see and that our brains reconstruct images by keeping track of those changes. It is true that the rods and cones in our retinas adapt rapidly to sustained activation, so any image that is fixed on any part of the retina will fade. However, when we "fixate" on some object of interest, the stability of our gaze is an illusion: our eyes are never still. "Fixational eye movements"—*microsaccades, ocular drifts, and ocular microtremor*—mean that whenever we look steadily at something, the retinal image is constantly "jittering," however still we perceive it to be. The fixational eye movements "work around" the problem of photobleaching to maintain a constant image of a constant scene. When lovers gaze into each other's eyes what they see is not an illusion, though what they read there might be. However, that constant image is encoded not in a stable collage of active retinal neurons, but in a constantly flickering and jittering map.

The properties of each neuron determine whether it will adapt to a sustained stimulus or maintain a constant response.[7] These properties determine the *pattern* of spikes that a neuron generates—the particular sequence of spikes that a neuron fires in response to a given stimulus. The sequence matters for many reasons, but especially because spikes that are clustered together release more neurotransmitter than the same number of spikes generated sparsely, and because the effects of packets of neurotransmitter released in quick succession can summate to give a stronger signal to the next neuron in the chain. Thus mechanisms that govern the generation of patterns of spiking activity are of major importance.

The first insight came from Herbert Gasser and Josef Engaler, who in 1944 were awarded the Nobel Prize in Physiology for showing that spikes in peripheral nerves were followed by slow "afterpotentials" that produced prolonged changes in excitability.[8] They saw that these properties would also be present at the site of spike initiation, implying that neurons might retain some "memory" of recent spike activity that would affect how they responded to synaptic input. We now know that, in different neurons, spike activity generates different sequences of afterpotentials that can generate complex patterns of spiking. *Hyperpolarizing afterpotentials* make neurons less likely to fire again after they have been active, while *depolarizing afterpotentials* make them more likely to fire again. In some neurons, a flurry of inputs that lasts just a second might trigger a burst of spikes that lasts longer than a minute. This might be useful, as a way of holding for a while the memory of a transient event, but when a neuron responds to an input by a stereotyped burst of spikes, any information that was present in the fine structure of the input is irretrievably lost.

Whenever a signal passes from one neuron to another, information is lost. A sensory cell might respond to a stimulus by a fluctuating electrical signal that reflects the sensory signal, but when that signal is transformed into a train of spikes, the details of the fluctuations are lost. When the train reaches a synapse, it is converted to another fluctuating chemical signal, and this has an inconsistent relationship to the train of spikes that evoked it: it depends on the availability of vesicles for release, which fluctuates stochastically. In the postsynaptic cell, that chemical signal activates receptors and is transformed again into a fluctuating electrical signal with further loss of fidelity. That new signal, along with signals from thousands of other synapses, triggers a new train of spikes—and the patterning of that new train is influenced by the intrinsic properties of the neuron. Some neurons generate bursts: short fast bursts, long slow bursts, complex fractured bursts—bursts occur in different neurons with a seemingly endless variety. Other neurons that have a "pacemaker potential" fire with a metronomic regularity; many others fire seemingly at random. The spiking activity of neurons in the brain has an inconsistent, gossamer-like association with information in the sense in which we would normally understand it.

If connections between neurons were all purposeful and all activity were meaningful, then this vast network would indeed have a massive capacity for processing information. The human brain has at least 80 billion neurons, each of which make, on average, 10,000 connections with other neurons—8×10^{14} points of information transfer. Many neurons can fire 200 spikes per second, although few do so except rarely and briefly. But if each neuron *might* fire a spike every 5 milliseconds, the rate of information processing might seem to be between 10^{16} and 10^{17} calculations per second.

At the rate at which the power of computers is increasing, desktop computers might soon do better than this. Donald Moore pointed out in 1965 that, every year since the integrated circuit was invented, the number of transistors per square inch had doubled, and he predicted that this trend would continue—a prediction that became known as Moore's law.[9] In 1965, chips had about 60 transistors; by 2014, the 15-core Xeon Ivy Bridge-EX had more than 4 billion. Computational capacity is measured in FLOPS—floating point calculations per second. In 1997, 10^9 FLOPS cost about $42,000; by 2012 this had fallen to about 6 cents, and, in 2016, a supercomputer in China became the fastest in the world at 10^{17} FLOPS. If costs continue to fall and power continues to increase at present rates, $1,000 will buy 2×10^{17} calculations per second by 2023, and desktop computers will apparently have a computational capacity equivalent to that of the human brain.

However, this calculation of the brain's computational power is misleading. The 10,000 outputs of a neuron are not independent: they all carry the same information, and they don't go to 10,000 different neurons but usually to just a few hundred at most. Synapses are unreliable, neurons are typically noisy and erratic, and most neurons never fire at anything like 200 spikes per second—indeed, many are not capable of doing so. The brain also has extensive redundancy: Parkinson's disease has few symptoms until about 85% of the neurons in the substantia nigra have degenerated. Spiking activity in the brain has nothing like the computational capacity claimed for it.

The brain determines how our bodies will respond to changes in our environment, and *all* of its neurons contribute, to some extent, to the generation of *all* behaviors. Exactly how they contribute depends on their intrinsic properties, which are heterogeneous, and on their connections with other neurons, which are part rule-governed, part molded by the contingencies of experience, and part accidental. The agents within our brain are messy and unreliable products of genetics, epigenetics, accident, and chance. This makes any analogy between digital computers and the brain misleading.

The era of digital computers might soon be closing. Transistors are built from thin wafers of silicon that form logic gates by allowing a current to pass if a small voltage is applied to the wafer but not otherwise. If the wafers are too thin, "electron tunneling" will occur: a current will erratically find its way through even without any applied voltage. We are fast approaching this hard limit to miniaturization.

On the horizon, quantum computing promises vast increases in computing power. Whereas conventional logic gates have two states, "true" or "false," quantum devices have many possible states, and, because they are built on a molecular scale, the density of packing of devices is practically unlimited. Quantum devices will be noisy and unreliable, and new computer architectures must be developed that are self-organizing, fault-tolerant, and self-correcting, and that function robustly in the presence of noise. New software must find ways of coping with uncertainty in the operation of individual elements. These are characteristics of the brain, which should be a source of inspiration for new architectures and systems: while computers might soon be powerful enough to help us understand the brain, we might have to understand the brain better before we can program them. This prospect challenges the pretensions of neuroscience: do we understand much, or do we just know a lot? Is our understanding of the brain anything more than a lazy borrowing from a (possibly mistaken) understanding of how digital computers process information?

The classical view is a monumental edifice, a monumental achievement. Artificial neural networks, constructed by analogy with this classical view,[10] drove progress in artificial intelligence and machine learning with wide applications in data analysis and robotics. Such networks are magnificent calculators, exquisite rational agents. But where in this edifice are our *passions*—where is the "heart" of the brain?

Shelley's poem "Ozymandias" ends,

And on the pedestal these words appear:
"My name is Ozymandias, king of kings:
Look on my works, ye Mighty, and despair!"
Nothing beside remains. Round the decay
Of that colossal wreck, boundless and bare,
The lone and level sands stretch far away.

4 Enlightenment

Reason is, and ought only to be the slave of the passions, and can never pretend to any other office than to serve and obey them.
—David Hume (1711–1776)[1]

Our scientific method has no universal rule book; it is an uneasy coalition of divergent philosophies. On the one hand is the reductionist manifesto, expressed by Thomas Hobbes in the preface to *De Cive* in 1651:[2]

Everything is best understood by its constitutive causes. For, as in a watch or some such small engine, the matter, figure and motion of the wheels cannot well be known except it be taken asunder and viewed in parts.

Hobbes saw the mathematical certainty of geometry as that to which science should aspire. He viewed the purpose of science as being to establish *causal explanations*—deductions from observations to necessary conclusions. If we observe, for instance, the spiking behavior of a neuron, the reductionist invites us to ask *why* it behaves in that way. This question carries in two directions: what *caused* that behavior, and what are its *consequences*; and the reductionist seeks deterministic accounts of both.

On the other hand is the vision of science expressed by Karl Popper in *Conjectures and Refutation* (1963). This vision denied that absolute certainty can come by inductive reasoning from observations: there are always different ways of "explaining" observations, and science must find ways of choosing between them. We want explanations that are simple and elegant yet powerful—theories that imply more than we can know from the mere accumulation of facts. Such theories must forever be provisional: in predicting things that we do not yet know, they are forever open to disproof.

There is no more rational procedure than the method of trial and error—of conjecture and refutation: of boldly proposing theories; of trying our best to show that these are erroneous; and of accepting them tentatively if our critical efforts are unsuccessful.[3]

However, we cannot advance by always and forever questioning the foundations of our understanding. As Popper put it:

Science does not rest upon solid bedrock. The bold structure of its theories rises, as it were, above a swamp. It is like a building erected on piles. The piles are driven down from above into the swamp, but not down to any natural or "given" base; and if we stop driving the piles deeper, it is not because we have reached firm ground. We simply stop when we are satisfied that the piles are firm enough to carry the structure, at least for the time being.[4]

Neuroscience progresses through an uneasy combination of reductionism and the "leaping imagination" of Popperian science, and through *discovery*. New discoveries often arise from reductionist approaches. They often depend on technological advances: from molecular biology, in its ability to dissect, transform, and reconstruct the basic functions of cells; from physics, in new ways of imaging cells and the molecules and organelles within them; from mathematics, harnessing vast computational power in the analysis of structures and networks; or from chemistry, in the ability to manipulate the basic elements of neural signaling. Ideas are often inspired by discoveries. Discoveries about mechanisms in cells usually lead to some ideas that we can test by the same means that we employed to make those discoveries. But to understand how mechanisms in cells construct the behavior of complex systems comprising many interconnected cells is harder, and we must find ways to *test* our ideas, not just ways to advertise their possibility.

My home of Edinburgh was, at the close of the eighteenth century, the hub of the Scottish Enlightenment. It was home to Joseph Black, the founder of thermochemistry, who discovered carbon dioxide and magnesium; to his student Daniel Rutherford, who discovered nitrogen; to the geologist James Hutton, whose work revealed the deep time through which the world has evolved; to Adam Ferguson, the "father of modern sociology"; to Lord Monboddo, the founder of comparative linguistics; to Adam Smith, the author of *The Wealth of Nations*; to Colin Macfarquhar and Andrew Bell, who founded the *Encyclopaedia Britannica*; to Alexander Monro (primus), who founded the Medical School at Edinburgh and rescued anatomy from the dogma of classical authority; and to his son (secundus), who discovered the lymphatic system. Another native son was William Cullen, who recognized the importance of the mind in healing and the placebo effect, and who coined the term *neurosis*. His lectures and textbooks became famous internationally and grew the size and influence of the Medical School. One of his American students was inspired to name his son after him: William Cullen Bryant became the "first American poet."

Friend to them all was the philosopher David Hume, whose skeptical approach led a later son of Edinburgh, the novelist Robert Louis Stevenson, to declare that Hume had "ruined philosophy and faith."[5] In his *Enquiry Concerning Human Understanding* (1748) Hume argued that any reasoning from mere observations is undermined both by the fallibility of our senses, on which all observations of fact depend, and by the logical inability to deduce causality from association however consistently two events might be associated in time.[6] However often we have seen that day follows night, we cannot be certain that the sun will rise tomorrow, nor can we safely assume that it is darkness that causes the sun to rise.

This is not because the laws of nature might change while we sleep, but because we might have mistaken the laws by our inferences from incomplete evidence. We can never prove the truth of any theory, no matter how much evidence we might have assembled: tomorrow might yet bring a counterexample. Scientists therefore avoid talking of proof, almost as though it were unlucky to do so.

But although we can never prove a causal association we might yet *disprove* one, and Karl Popper's answer to Hume's "problem of induction"[6] was that the scientific method must exploit this. In *The Logic of Scientific Discovery* he argued that we make progress only with the aid of imagination.[4] From observations we construct hypotheses that we test by experiments, aiming not to verify them but to refute them. We develop our understanding by culling refuted theories, leaving only those that, for now, have withstood determined challenges. Popper's vision has been embedded in the ethos of academic journals, research funders, and grant-awarding committees, all of which yearn for innovative hypotheses and "killer," hypothesis-destroying experiments by which they can be tested.

If we observe the behavior of a neuron, the reductionist manifesto invites us to ask what caused that behavior, and what its consequences are. Here is the rub. We can speculate about these things, but as Hume put it, "causes and effects are discoverable, not by reason but by experience."[7]

We gain experience through experiments, and from this we try to formulate hypotheses that can be tested. Often the only falsifiable hypotheses we can make are rather trivial: about *immediate causes* of a behavior of a particular neuron or about its *direct consequences*. We really want bold hypotheses, hypotheses about the meaning of its behavior for the whole organism—hypotheses about its *physiological significance*. But the behavior of any one neuron in a mouse or a man has *no* physiological significance. Every neuron receives thousands of inputs from hundreds of neurons and sends outputs to hundreds of others. Functions are enacted not by neurons acting alone but by populations acting on other populations.

If any neuron were fully representative of a population commonly engaged in a single function, then we might, by observing one neuron, glimpse that collective behavior. But all neurons are individuals, with individual quirks and eccentricities.

I, like you, am a human being. We share the same genes, with a few minor differences, but those did not solely determine who we are now. The environment into which we are born, our early life experiences and our interactions with close kin, define us also. Particular experiences might seem fleeting and ephemeral, yet they leave their mark on our behavior. These things that are true of you and me are true of any two neurons, however similar their genetic and developmental fates. As we will see, adjacent neurons serving a common physiological function can respond differently to the same stimulus. No neuron receives the same inputs as even its closest neighbor, or expresses exactly the same genes, or has the same morphology or the same intrinsic properties.

Our understanding of the brain is built of boxes inside boxes inside boxes inside boxes. The largest box contains about 20,000 genes and the hundreds of thousands of proteins that they encode. Some who rummage in this box are concerned with the structure of complex molecules and how the shape of a receptor enables a peptide to bind with it. Some worry about how the genes are regulated, what proteins bind to their regulatory regions, and how this affects the level of expression. Some study how proteins are cleaved by enzymes, and what chemical reactions occur. The next box contains the many different types of neurons. Here, the puzzles include understanding the properties of membranes that allow neurons to generate patterns of spike activity, the intracellular signaling pathways that are activated by signals received, the mechanisms by which activity is coupled to secretion and gene expression, and how all of these are influenced by physiological signals. The next box contains the clans of neurons, and here the questions are about how each clan is organized, how cells in a clan communicate with each other and with other clans. This box is about the *networks* of neurons, and how information flows through them. The smallest box of all is about *systems*: how the networks of clans achieve important ends, like maintaining a constant body weight through a lifetime of changing habits of eating and exercise, like delivering a baby, like recognizing a friend, like deciding when to run and when to fight. These boxes must fit together, but often it's hard to see quite how they do. Although we think with our brains and through the stuff of our neurons, it is not clear that by studying neurons we can say much about how we think. The classical foundations of our understanding of the brain have served us well, but the question must arise as to whether these foundations are sound enough to explain the behavior of neuronal systems.

To understand the enactment of function we must explore the behavior of populations of neurons that act collectively to influence other populations. We must know what neurons in a clan have in common and what differences separate them from other clans. We must understand the interactions among neurons of a clan, and theirs with neighboring clans, and how this aggregated network responds to inputs from other networks of clans. We must also know a lot about the functions that the clans enact before we can formulate testable ideas about *how* those functions are enacted.

Among the advantages of neuroendocrine systems are that we can put names to the neurons we study. Vasopressin neurons do "what it says on the tin": they secrete vasopressin, which acts on the kidneys to concentrate the urine. We can measure what is secreted and study its consequences. We can study the neurons in exquisite detail, and on a scale by which we can infer the behavior and properties of the population. Neuroendocrine systems are tractable, amenable to reductionist interrogation, and this enables us to build an understanding that can generate the bold hypotheses we seek. They are also important, if we accept that our health and survival are important. Understanding them must underpin our assault on obesity and diabetes, stress and depression, hypertension and infertility, and disorders of social and sexual behavior. I wouldn't argue that we should study only systems that seem important in this narrow sense. When I lose my keys, as I often do, it might seem sensible to look in obvious places, but if they were in obvious places they wouldn't be lost. It is often better not to look for your keys at all, but to do something else that leads you to different places, where you might find your keys or something much better.

Each neuroendocrine system is different from every other. Will anything we learn be relevant to the rest of the brain? I think so, for two reasons, both rooted in Hume's *Enquiry*.

> The utmost effort of human reason is to reduce the principles, productive of natural phenomena, to a greater simplicity, and to resolve the many particular effects into a few general causes, by means of reasonings from analogy, experience, and observation.[7]

Karl Popper, in *The Logic of Scientific Discovery*, emphasized the importance of imagination: "Bold ideas, unjustified anticipations, and speculative thought, are our only means for interpreting nature: our only organon, our only instrument, for grasping her."[8] Where do these ideas come from? They come, said Hume, not from our reason but from our experience, and they come as *analogies*. We see patterns in clouds and stars, emotion in abstract images, messages in parables, and metaphor in poetry. The oxytocin and vasopressin systems are important "model systems" in neuroscience not because they have revealed universal truths, but by inspiring hypotheses that might or might not hold elsewhere in the brain.

My second reason is that the hypothalamus affects our emotions and behavior. Some populations of neurons determine not only how much we eat and drink but also what and when we eat and drink. Others drive daily rhythms of drowsiness and wakefulness, others control aggression, others sexual behavior, others love and friendship, others our moods and our responses to stress and threat.

If we ask someone why they did some particular thing, as likely as not they will give a reason that is a "rationale" for their actions. But how much of our behavior is really governed by reason? How many reasons are mere rationalizations of behavior driven by passions, biases, and instincts unacknowledged? I'll give Hume the last word:

> This operation of the mind, by which we infer like effects from like causes, and vice versa, is so essential to the subsistence of all human creatures, it is not probable, that it could be trusted to the fallacious deductions of our reason, which is slow in its operations; appears not, in any degree, during the first years of infancy; and at best is, in every age and period of human life, extremely liable to error and mistake. It is more conformable to the ordinary wisdom of nature to secure so necessary an act of the mind, by some instinct or mechanical tendency, which may be infallible in its operations, may discover itself at the first appearance of life and thought, and may be independent of all the laboured deductions of the understanding.[9]

5 Far from the Madding Crowd

The thin grasses, more or less coating the hill, were touched by the wind in breezes of differing powers, and almost of differing natures—one rubbing the blades heavily, another raking them piercingly, another brushing them like a soft broom. The instinctive act of humankind was to stand and listen, and learn how the trees on the right and the trees on the left wailed or chaunted to each other in the regular antiphonies of a cathedral choir; how hedges and other shapes to leeward then caught the note, lowering it to the tenderest sob; and how the hurrying gust then plunged into the south, to be heard no more. The sky was clear—remarkably clear—and the twinkling of all the stars seemed to be but throbs of one body, timed by a common pulse.

—Thomas Hardy (1840–1928), *Far from the Madding Crowd*

Neuroendocrinology is about things that matter for our health and well-being: stress and appetite, metabolism, body rhythms, growth, and all aspects of reproduction: the reproductive cycle, sexual behavior, pregnancy and parturition, lactation and maternal behavior. In the hypothalamus, specialized neurons control the secretion of pituitary hormones that act on the gonads, the thyroid and adrenal glands, the kidneys, liver, heart and gut. These neuroendocrine systems offer a window on the brain. Their products can be measured relatively easily and, with wit and ingenuity, their effects are determinable. The activity of these neurons is interpretable to a degree equaled in few other areas of neuroscience.

If we ask of any neuron in the brain, what does it really *do*, the answer is often frustratingly incomplete. Even if we know how it responds to stimuli, what it makes and to which other neurons it talks, we might yet not know what it does that matters. But we can know much of what, for example, vasopressin and oxytocin neurons do even before we know how they do it. They send axons to the posterior pituitary gland, and what they secrete from there can be measured in the blood and has demonstrable consequences for important physiological functions. Vasopressin regulates the kidneys, and oxytocin regulates milk letdown from the mammary gland and controls uterine contractions during parturition—we now know that they both do much, much more than this, but when I started out as a neuroendocrinologist we knew just these things for sure.

These neurons once seemed unusual in that they use peptides for signaling. Peptides are much larger than conventional neurotransmitters, and hard to make. A peptide is a chain of amino acids, and which particular amino acids and in which order they are assembled is determined by the gene for that peptide. Oxytocin has nine amino acids and a molecular weight of about 1,000 daltons, and the oxytocin gene must produce it as a fragment of a much larger *precursor peptide,* which has a molecular weight of about 23,000 daltons. The precursor contains the oxytocin sequence and other sequences that determine what the cell will do with the oxytocin. Part of the precursor determines that it will be packaged into vesicles to be transported to sites where the vesicles can be released. Another part is important for folding the precursor in a way that enables it to aggregate with other precursor molecules so they can be packaged compactly in a vesicle. The vesicles also contain enzymes that cleave oxytocin from the rest of the precursor. It's complicated and expensive, but the final product is a powerful molecule: oxytocin will survive in the extracellular fluid for much longer than ordinary neurotransmitters, it can act on cells at much lower concentrations, and it acts at sophisticated receptors that have a range of properties through which they control complex signaling pathways within those target cells.

When I began as a neuroscientist, oxytocin was one of only a handful of peptides known to be released by neurons. Now, peptides are no longer rare and curious: more than a hundred are so far known to be released by neurons, to act not only on other neurons but also on the even vaster numbers of nonneuronal cells in the brain. Some peptides control the blood flow that carries the oxygen that neurons consume. Others control the growth of synapses between neurons and the removal of disused synapses. Many regulate the expression of genes in the cells that they target. Some control the shape as well as the function of the glial cells that are so abundant in the brain, which cleanse the external environment and provide growth factors and metabolites to support neuronal function. Many influence neuronal excitability directly, or modulate the effects of neurotransmitters.

One striking characteristic of peptides is their ability to orchestrate behavior—to coordinate different systems to evoke a coherent, adaptive, organismal response, be it maternal behavior, aggression, sexual arousal, or behaviors associated with hunger and thirst: foraging, feeding, and drinking. An injection of NPY (*neuropeptide Y*) into the brain will provoke feeding, and an injection of α-MSH (*α-melanocyte-stimulating hormone*) will stop it. Orexin can wake a sleeping beauty. Angiotensin can make a horse drink. Vasopressin can make *her* turn to violence, and oxytocin make *him* maternal. The ability of peptides to evoke such dramatic "global" consequences invites us to look for things that are special about peptide signaling, and for this, the neuroendocrine systems have been a rich source of inspiration.

The hypothalamus is at the base of the brain. If you curl your tongue back as far as you can and press it on the roof of your mouth, the hypothalamus will be almost on the tip of your tongue. Your pituitary gland sits just a little farther back, below the hypothalamus and connected to it by the *neural stalk*, a thin strip of tissue that contains blood vessels and a bundle of axons. The gland in humans is described in Wikipedia as being the size of a pea. So common is this description that it seemed likely to be wrong, as I confirmed by examining a selection of peas. Wikipedia also gives the weight of the human pituitary as about half a gram, and in this it is more correct. The pituitary in a human is at least five times the average size of my peas; it is more like the size of a chickpea.

The pituitary has three distinctive lobes that produce nine main hormones, and the anterior lobe makes six of these. *Luteinizing hormone* (LH) and *follicle-stimulating hormone* (FSH) are gonadotropins, and regulate the gonads—the ovaries and testes. *Prolactin* regulates the production of milk by the mammary gland. *Growth hormone* acts on the liver to stimulate the production of insulin-like growth factor 1, which promotes bone growth and muscle development. *Thyroid-stimulating hormone* regulates the thyroid gland, which controls our metabolic rate by secreting other hormones that act on almost all of the cells in our body, and which are also critically important for brain development. *Adrenocorticotropic hormone* (ACTH), secreted in response to stressors, regulates the production of glucocorticoid hormones by the adrenal gland, and these affect blood glucose levels, fat metabolism, and blood flow to muscles. These six pituitary hormones are made in five types of endocrine cells: LH and FSH are both made by one of these cell types, the *gonadotrophs*. The gonadotrophs, lactotrophs, somatotrophs, thyrotrophs, and corticotrophs are all regulated by factors released from hypothalamic neurons into blood vessels at the base of the brain that communicate between the hypothalamus and the pituitary. This was first shown by Geoffrey Harris, who, by transplanting the pituitary gland to various sites in the body, and by severing the blood supply to it from the hypothalamus, showed that the pituitary could survive without its blood supply from the hypothalamus, but could not function.[1]

The intermediate lobe of the pituitary contains cells that contain α-MSH, which promotes the production of the pigment *melanin* in the skin. These cells (melanotrophs) are directly innervated by axons from the hypothalamus—as are endocrine cells in the anterior pituitary of fish.

The posterior lobe of the pituitary (also called the *neural lobe* or *neurohypophysis*) is also innervated by neurons of the hypothalamus, but it has no endocrine cells: the axonal endings of these neurons secrete oxytocin and vasopressin directly into the systemic circulation.[2] This system has been the focus of intense interest for many years, because of its unique tractability. The neurons are large, and they are aggregated in compact nuclei in the hypothalamus, making them relatively easy to recognize and study. They make huge amounts of their product—enough to be active throughout the body, making them amenable to studying the regulation of peptide synthesis, metabolism, and transport. Their axons are compactly and conveniently aggregated in the neural lobe away from any other neuronal elements, making it possible to study their properties and how electrical activity in them regulates secretion. And what they secrete is readily measurable in the blood and in the brain and can be related directly to physiological functions (figure 5.1).

Neuroendocrinology has its icons, and the *milk-ejection reflex* is one. *The Origin of the Milky Way*, painted by Tintoretto in about 1575, hangs in the National Gallery in London; it shows Jupiter holding the infant Hercules to Juno's left breast. Some milk from that breast spurts up to form the Milky Way; but milk also spurts from the unsuckled breast. From this we might correctly conclude that, while the letdown of milk is a response to the suckling of an infant, this response is not local to the breast that is suckled—it involves a systemic mediator. That mediator is oxytocin, and it causes cells of the mammary gland to secrete milk into a collecting duct. Oxytocin is secreted into the blood from the posterior pituitary but is *made* in the hypothalamus, in the magnocellular neurons of the supraoptic and

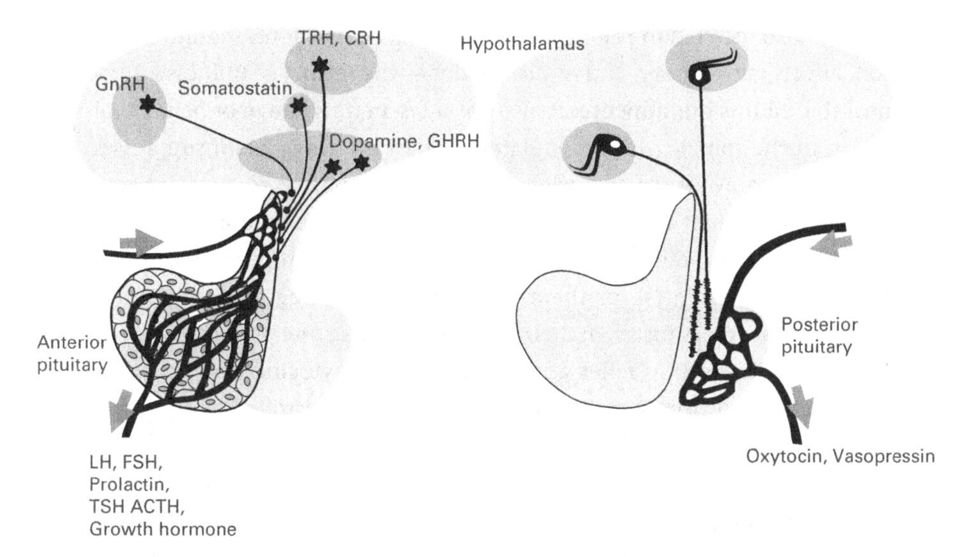

Figure 5.1
The neuroendocrine systems of the hypothalamus. The anterior pituitary contains endocrine cells that produce six major hormones. The secretion of these hormones is controlled by "hypothalamic hormones" released by small populations of neurons in different regions of the hypothalamus. LH and FSH secretion is controlled by GnRH from neurons in the rostral hypothalamus; growth hormone secretion by GHRH from the arcuate nucleus and by somatostatin from the periventricular nucleus; prolactin secretion by dopamine from the arcuate nucleus; TSH secretion by TRH from the paraventricular nucleus; and ACTH secretion by CRF, also from the paraventricular nucleus. The posterior pituitary contains no endocrine cells: here, the two hormones oxytocin and vasopressin are secreted from the axonal endings of magnocellular neurons of the supraoptic and paraventricular nuclei. Between the posterior and anterior lobes is the intermediate lobe (not shown here). It contains cells that produce MSH, and these are directly innervated by another population of dopamine cells in the arcuate nucleus.

paraventricular nuclei. When young suckle, oxytocin is secreted as a pulse into the blood; that pulse is the result of nearly synchronous activation of every magnocellular oxytocin neuron in an intense burst of spike activity that lasts for little more than a second.

Oxytocin has many roles. It is secreted during sexual activity: it modulates the lordosis reflex, the posture that in many female mammals is the signature of their willingness to mate; it promotes penile erection in males; and it is secreted during parturition, with potent effects on the uterus. In rats, it is also secreted in response to food intake: it promotes sodium excretion, affects gut motility, and induces satiety—the feeling of fullness after a meal that blunts our appetite. It also has a fascinating range of behavioral effects: in the spinal cord it modulates pain, in the amygdala it suppresses fear, in the paraventricular nucleus it dampens responses to stressors, and best known of all, it influences social attachment. Mice that lack oxytocin can still mate, get pregnant, and give birth; oxytocin is important for these things but not essential for them. However, their young will not survive unless they are cross-fostered; their mother will nurse them but cannot feed them: the milk-ejection reflex absolutely requires oxytocin.[3]

As well as showing that the anterior pituitary is controlled by hypothalamic "releasing factors," Geoffrey Harris helped build our understanding of the posterior pituitary. He knew that oxytocin could stimulate milk letdown in lactating animals, he saw that this could be measured as an increase in pressure inside the mammary gland, and, with Barry Cross, he showed that electrical stimulation of the neural stalk increased intramammary pressure. Harris and Cross proposed that, in response to suckling, oxytocin is secreted as a result of an increase in the spiking activity of hypothalamic neurons, propagated along the axons that pass through the neural stalk.[4,5] When Barry Cross moved to the Chair of Anatomy at Bristol, he set out to test this by recording the spiking activity of oxytocin neurons using microelectrodes.

The supraoptic nucleus of the hypothalamus contains *only* oxytocin and vasopressin neurons, all of which project to the posterior pituitary. However, this small site is surrounded by other neurons, and a microelectrode aimed there, however carefully, will not always hit its target. The breakthrough came from finding a way to be certain that the microelectrode has reached the intended target—from finding a way to *identify* the neuroendocrine neurons. In 1966 Kinji Yagi showed that electrical stimulation of the neural stalk would trigger spikes that were conducted not only "down" the axon, toward the nerve endings—but also "up" the axon (*antidromically*), toward the cell bodies. This made it possible to identify the neurons that project to the pituitary,[6] and thus to study them in living animals.

In Barry Cross's department, Dennis Lincoln and his PhD student Jon Wakerley set out to study the milk-ejection reflex.[7,8] In anesthetized lactating rats, they cannulated a mammary gland to record intramammary pressure and placed a stimulating electrode on the neural stalk. They then guided a microelectrode into the supraoptic nucleus, using landmarks on the skull to determine the appropriate point of entry of the microelectrode into the brain. When they encountered a neuron, they could tell whether it was a supraoptic neuron by whether it displayed antidromic spikes when the neural stalk was stimulated.

After finding a supraoptic neuron, they allowed a litter of pups to suckle. The first surprise was not how strongly the suckling affected supraoptic neurons but how *little* effect it had. Some speeded up slightly, others slowed down, but generally nothing much happened. Some "phasic" neurons fired with long bursts separated by long silent periods; many others fired slowly and apparently randomly. After about 15 minutes of suckling, some of the slow-firing cells showed an explosive burst of spikes followed about ten seconds later by a sharp increase in intramammary pressure. These *milk-ejection bursts* recurred every few minutes while the pups were suckling, and each was accompanied by a *stretch reflex* in which the pups extended their forelegs stiffly to press on the mammary gland while they sucked.

Milk-ejection bursts last on average for just two seconds and contain about 100 spikes; within the first few spikes the bursts reach a peak of intensity that is sustained for about half a second, then wanes, and is followed by several seconds of silence. This pattern is stereotyped; the first bursts are smaller, but once the reflex is established, successive bursts in the same cell are very consistent in size and shape. The bursts occur at almost exactly the same time in all of the magnocellular oxytocin neurons. Other neurons in the supraoptic nucleus that did not participate in these bursts could be assumed to be vasopressin cells, including most of the phasic neurons.

Thus came the realization that, in response to suckling, oxytocin is secreted in pulses that result from synchronized bursts of spike activity. While the reflex is a response to the sucking of pups at the nipples, the activity of oxytocin cells does not passively follow this input. Rather, the input "permits" the oxytocin neurons to display bursts that are generated by some deterministic process. There is, however, something special about the suckling input: many other stimuli activate oxytocin cells strongly but never produce bursts.

This discovery—and it was a discovery, for although it may now seem that the nature of the reflex could have been predicted—came as a complete surprise. After these experiments, no study of any hormone secretion was complete without studying not just how much hormone was secreted but also the pattern in which it was secreted. Other workers began taking frequent samples to study hormone secretion and found that LH, FSH, prolactin, ACTH, thyroid-stimulating hormone, and growth hormone were also secreted in pulses. Two questions came to dominate neuroendocrinology: *why* are hormones secreted in pulses, and *how* are the pulses generated.

Geoffrey Harris had noted that electrical stimulation of the neural stalk would evoke an increase in intramammary pressure only if high frequencies of stimulation were used. This is partly because the response of the mammary gland has a narrow dynamic range. Oxytocin causes milk letdown by causing myoepithelial cells of the mammary gland to contract: but for any particular myoepithelial cell this is pretty much an all-or-nothing response. There must be enough oxytocin before it has any effect, and a maximum effect is achieved at a dose that is not much more than just enough.[9] In a lactating rat, the maximum effect is achieved when about 2 ng of oxytocin is injected intravenously as a bolus—the same amount as is secreted after each milk-ejection burst. If the same dose is given slowly it has little effect, and if much higher doses are infused slowly, then, after an initial peak the intramammary pressure diminishes even while oxytocin is still being infused, and the gland becomes insensitive to oxytocin. Thus the mammary gland *requires* pulsatile secretion of oxytocin.

Many organs and tissues desensitize when exposed to continuously high levels of a hormone. However, there are exceptions: in late pregnancy, oxytocin precipitates uterine contractions that continue in the presence of high levels of oxytocin. Moreover, in rats, oxytocin has another function; at low concentrations it stimulates sodium excretion (*natriuresis*), partly by an effect on the kidneys and partly by stimulating the secretion of a natriuretic hormone from the heart. This natriuretic effect requires continuous exposure to oxytocin, and the effect is graded according to concentration. Thus, whether pulses are necessary or not depends upon the properties of the target tissues, which implies that there has been co-evolution of the properties of the target organ and the patterning of hormone secretion.

There is another phenomenon behind Harris's observation. *How much* oxytocin is secreted by electrical stimulation depends on the frequency of stimulation: each spike within a burst releases much more oxytocin than the isolated spikes that occur between bursts.[10] The background activity of oxytocin neurons is about 2 spikes per second, and during a burst it rises to 50–100 spikes per second. But the bursts occur only every few minutes, last for only about two seconds, and are followed by up to ten seconds of silence. So, overall, suckling hardly increases the spiking activity at all; the reflex involves more a reorganization of activity than an amplification, and it is effective because bursts are so efficient at releasing oxytocin. Again, this is true but not inevitable. Stimulus-secretion coupling—the manner in which electrical activity affects secretion—is often nonlinear but the nonlinearities are different in different systems. The properties of the terminals and the mechanisms of burst generation have coevolved.

Oxytocin's role in milk ejection is indispensable: mothers that lack oxytocin cannot feed their young. By contrast, although oxytocin is named for its effects on uterine contractility (from the Greek for "quick birth"), mice that lack oxytocin deliver their young relatively normally. In 1941 James Ferguson in Toronto reported that distension of the uterus and cervix could induce oxytocin secretion in the pregnant rabbit: this *Ferguson reflex* is now known to be a feature of all mammals, and since oxytocin can stimulate uterine contractions, which in turn stimulate oxytocin secretion, it is a rare example of a biological positive feedback system.[11] Nevertheless, in the absence of oxytocin, other mechanisms can ensure a successful delivery. In the year that Ferguson described his reflex, Dey and colleagues reported on the effects of lesions to the neural stalk in pregnant guinea pigs that prevent any secretion of oxytocin.[12] Of 16 labors that they studied, 10 were prolonged and difficult but 6 were normal.

Geoffrey Harris had shown that electrical stimulation of the neural stalk could evoke uterine contractions in pregnant rabbits, but being a cautious and skeptical scientist, he remained unsure whether this meant that oxytocin was really secreted during parturition, or whether it was merely a "pharmacological" effect, of no physiological significance. His experiments prompted a trainee obstetrician, Mavis Gunther, to write to the *British Medical Journal*.[13] She had attended labor in a woman who was still breast-feeding a previous child, and had noticed that, during each uterine contraction, beads of milk appeared at the nipples. Many factors were known that could elicit uterine contractions but only oxytocin was known to induce milk letdown, so Gunther proposed that uterine contractions must have provoked the secretion of oxytocin, which therefore must act in a positive-feedback manner to provoke further contractions and thereby support parturition.

However, soon it came to be recognized that the placenta of pregnant women synthesized an enzyme, *oxytocinase*, that could degrade oxytocin, and that plasma concentrations of oxytocinase increased dramatically toward term.[14] This was strange: if oxytocin was important for parturition, why did the placenta produce an enzyme that destroys it?

In the 1980s, in Barry Cross's former department in Bristol, Alastair Summerlee and his colleagues recorded the electrical activity of oxytocin cells in conscious rats and rabbits throughout parturition and subsequent lactation.[15–18] The milk-ejection reflex in conscious rats was essentially identical to that described in the anesthetized rat, and similar bursts occurred during parturition; these were linked to the delivery of the young. In rabbits, things were only slightly different. Whereas rats allow their young to suckle continuously for long periods, rabbits nurse theirs for just a few minutes each day. Their oxytocin cells showed bursts that looked like those in rats, but each period of suckling was associated with several bursts in quick succession, and the bursts continued for a while after suckling, apparently in response to grooming of the nipples by the doe.

Parturition in both rats and rabbits was also accompanied by bursts that were associated with pulses of oxytocin secretion, and with delivery of the young. The recognition that oxytocin secretion was pulsatile during parturition cast a new light on the role of oxytocinase, for while high concentrations diminish the basal levels of oxytocin, they also "sharpen" pulses of oxytocin by reducing their half-life. By frequent blood sampling combined with methods to inactivate oxytocinase in the samples, Anna-Rita Fuchs and coworkers showed that spontaneous delivery in women is also accompanied by pulsatile secretion of oxytocin.[19]

But are the pulses *necessary*? This was less clear, as the uterus continues to contract in the continued presence of oxytocin. Nevertheless, in the rat, pulses are indeed a more effective way for oxytocin to drive parturition. In my lab at Babraham, Simon Luckman showed this by first interrupting parturition with morphine (a potent inhibitor of oxytocin neurons).[20] He showed that normal parturition could be reinstated with pulses of oxytocin given every ten minutes, but not by the same amount given by continuous infusion.

The trigger for parturition varies between species, but in all mammals oxytocin (or in marsupials, its ortholog, mesotocin) has a role.[21] Oxytocin is not essential—other mechanisms can compensate for its absence—but it is secreted in large amounts during labor, it acts on a uterus that expresses greatly increased numbers of oxytocin receptors, and blocking either oxytocin secretion or its actions slows parturition.

The explosive nature of milk-ejection bursts suggested that some positive feedback might be involved in generating them, and Barry Cross and his colleagues set about to test whether oxytocin itself could excite the oxytocin cells.[22] They recorded from magnocellular neurons in rats and rabbits and gave oxytocin either intravenously or directly to the neurons (by *iontophoresis*, which involves expelling drugs from a micropipette). The results were disconcerting: direct exposure to oxytocin excited many of the magnocellular neurons, but when given intravenously, even very large doses of oxytocin had no effect.

When these experiments were conducted there was no evidence that oxytocin was released in the brain, and Barry Cross recognized that the ineffectiveness of intravenous oxytocin meant that oxytocin secreted from the pituitary probably did not find its way back into the brain. He thus thought that the direct effects of oxytocin on oxytocin cells were of no physiological significance. He was right on the first issue: there is a blood-brain barrier to oxytocin, later demonstrated compellingly by Wim Mens and his colleagues in Utrecht, who injected massive amounts of oxytocin intravenously into rats and showed that, although plasma concentrations increased a thousandfold, levels in the cerebrospinal fluid scarcely changed.[23] But Cross was wrong on the second issue.

This became apparent when Philippe Richard and Marie-José Freund-Mercier in Strasbourg showed that small amounts of oxytocin injected into the brain dramatically facilitated the milk-ejection reflex.[24] While it might have been expected that oxytocin would excite oxytocin neurons, this is *not* what they saw. Oxytocin had no effect in rats that were not being suckled and only slight effects on the background activity of oxytocin cells in rats that were. But, for 15 to 30 minutes after injecting oxytocin, they saw an increase in the size and frequency of milk-ejection bursts. These effects could be evoked by injecting as little as 12 *picograms* into the brain—a tiny amount (figure 5.2).

The milk-ejection reflex had seemed to be a product of a sophisticated neuronal network that transformed a fluctuating sensory input (from the suckling of hungry young) into stereotyped bursts that were synchronized among the entire population of oxytocin cells. Yet injecting oxytocin

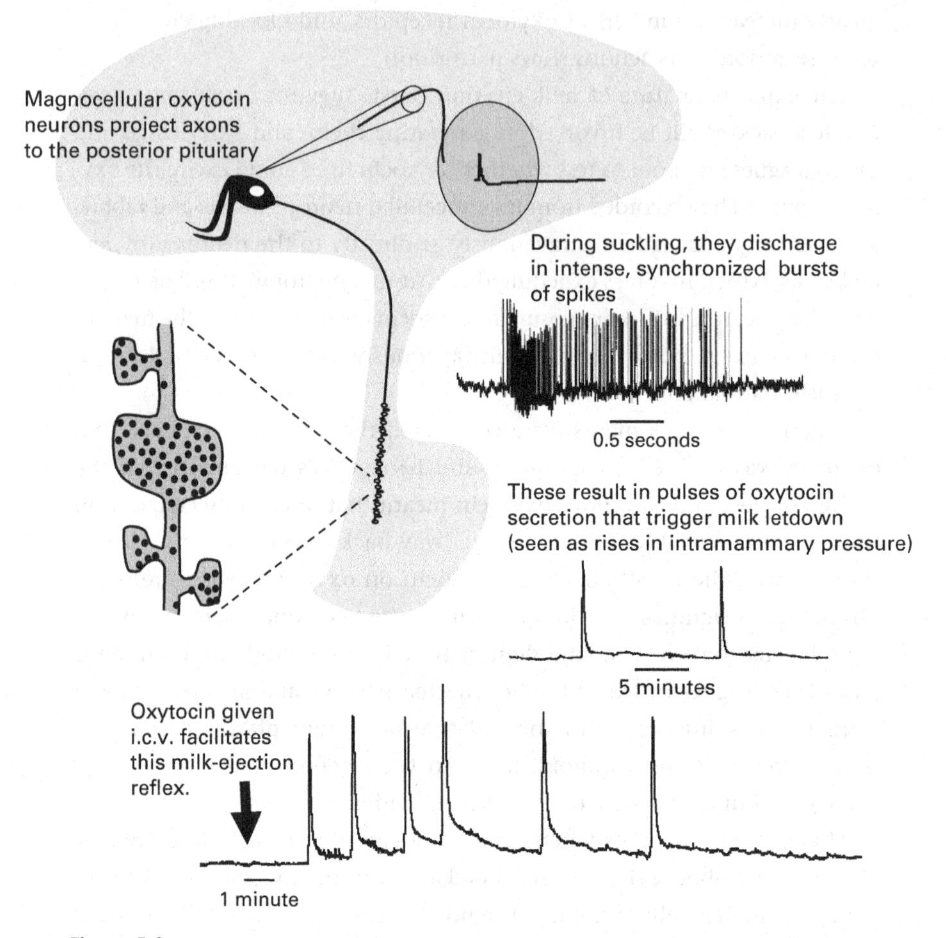

Figure 5.2

The milk-ejection reflex, uncovered by Wakerley and Lincoln using electrophysiological studies in anesthetized rats. In response to suckling, oxytocin cells discharge intermittently in brief synchronized bursts that evoke secretion of pulses of oxytocin, and these pulses induce abrupt episodes of milk letdown, reflected by increases in intramammary pressure. This reflex is dramatically potentiated by injections of very small amounts of oxytocin into the brain. This was the first clear demonstration that peptides, released centrally, can have potent physiological actions.

into the brain with no spatial or temporal sophistication resulted in an apparently specific modulation of this reflex, and did so at impressively low doses.

Did these observations reflect an action of oxytocin on oxytocin cells? Oxytocin produced simultaneous bursts, but the neurons do not communicate with each other either by chemical synapses or by electrical synapses. There seems to be no physical connection between the neurons in one supraoptic nucleus and those in the other, which are separated by the third cerebral ventricle. If there *is* positive feedback, why weren't bursts triggered by *any* stimulus that excited oxytocin cells? Bursts occurred during suckling and also during parturition, but never in response to any of a growing catalog of other kinds of stimuli that could excite oxytocin cells.

And yet, further experiments of Philippe Richard, Françoise Moos, and their colleagues showed that the effects of oxytocin on the milk-ejection reflex must reflect an action on oxytocin cells.[25] First, they used the technique of *push-pull perfusion* to sample fluid from the supraoptic nucleus during suckling. This involves placing a probe into the brain that consists of two concentric cannulae, one to slowly infuse artificial cerebrospinal fluid, and one to collect fluid at the same rate. Substances released in that region of the brain would mix with the infused fluid and be recovered by the collecting cannula. The researchers saw that oxytocin release was increased even before any increase in the blood was detected, and hence before the oxytocin cells showed any increase in spiking activity. Second, they found that small amounts of an oxytocin antagonist microinjected into just one supraoptic nucleus would block the reflex. To a physiologist, experiments with antagonists have a special significance: they reveal the actions of an agonist produced by the brain itself—an *endogenous* agonist. Thus oxytocin release in the hypothalamus was essential, and there had to be some communication between the nuclei if an intervention in one could block bursts in the other. Third, they studied isolated oxytocin cells in vitro and used a probe to measure the intracellular calcium concentration—a dye that produces a fluorescent signal when the calcium concentration increases. With this, they showed that oxytocin indeed acted directly on oxytocin cells: the oxytocin cells responded by releasing calcium from their intracellular stores.

These experiments were the first convincing demonstration of a physiological role for any peptide in the brain. They did not explain the milk-ejection reflex but defined the questions that had to be answered before it could be explained. Building that explanation took another twenty years. The questions posed had no precedent in our understanding. Where did the oxytocin that was released in the supraoptic nucleus come from, if not from synapses? What triggered its release, if that release was not governed by spiking activity? What synchronized the oxytocin cells, if they were not linked by either synapses or electrical junctions?

We now know that more than a hundred neuropeptides are expressed in different neuronal populations, and that most if not all neurons in the entire brain release one or more peptide messengers as well as a conventional neurotransmitter. Because peptides have a long half-life and act at receptors at very low concentrations, their actions are not confined to targets adjacent to the site of release. Importantly, peptides in the brain often have organizational and activational roles that seem more like the roles of hormones in the periphery.

Whereas conventional neurotransmitters are packaged in small vesicles that are released only at synapses, peptides are packaged in large vesicles in all parts of a neuron, including in the cell body and dendrites, and can be released from any of these. Peptides can be released by spiking activity, but, whereas at a synapse one spike might release one synaptic vesicle, it may take many hundreds of spikes to release one peptide-containing vesicle. Some stimuli can alter the availability of vesicles for release, and this changes the functional connectivity of neurons. Some stimuli, by mobilizing intracellular signaling cascades, can cause peptide release from dendrites without any release of conventional transmitters from axon terminals—which lack local calcium stores. Peptide signals can be long-lasting and can act at considerable distances from their site of release. And some neurons can change their phenotype in different physiological states, expressing different peptides.

This view of the brain is different from the conventional portrait. It shows the hypothalamus as a "Europe" of the brain, a confusion of small nations, each noisy, heterogeneous, and sometimes strident. Each nation contains multiple clans that use a variety of languages and other signals that act at diverse spatial and temporal scales to communicate with other clans and neuronal nations.

6 Pulsatile Secretion

Let us roll all our strength and all
Our sweetness up into one ball,
And tear our pleasures with rough strife
Through the iron gates of life:
Thus, though we cannot make our sun
Stand still, yet we will make him run.
—Andrew Marvell (1621–1678), "To His Coy Mistress"

The findings of Wakerley and Lincoln marked a watershed for endocrinology. Before 1970, it had been assumed that hormonal concentrations changed only sluggishly, so a single blood sample would characterize the "state" of any particular endocrine system. Just a few years later, no hormone system could be considered as understood until the *pattern* of secretion was characterized. Of all the endocrine systems, the gonadotropins displayed the most vivid, dynamic behavior. In females of all mammals, LH and FSH are secreted in large intermittent pulses whose frequency and amplitudes vary over the course of the ovarian cycle. At the midpoint of the cycle, the pulses fade and a massive surge of LH and FSH secretion triggers ovulation, heralding a dramatic change in behavior. At this time, females of most mammal species are not just receptive to sexual advances from males, but actively seek them.

LH and FSH are both made in gonadotroph cells of the anterior pituitary, but they are packaged in separate vesicles and secreted semi-independently. Through their control of the ovaries, these two gonadotropins orchestrate all of the physiological, hormonal, and behavioral events that lead to ovulation, mating, and reproduction. In males, FSH is required for spermatogenesis and LH for the production of the male "sex hormone," testosterone. In females, FSH controls the growth and recruitment of immature ovarian follicles, and LH controls the production of estrogen by those follicles, which acts on the brain to regulate sexual behavior.

Pulsatile secretion of LH and FSH is regulated by gonadotropin-releasing hormone (GnRH); this is synthesized in the hypothalamus and released in pulses that are carried by portal blood vessels to the anterior pituitary. That sentence captures decades of work by hundreds of scientists and some remarkable technical achievements.

Studies of the pattern of hormone secretion depended on sensitive means for measuring hormones, which depended on the development of *radioimmunoassays*, for which Rosalind Yalow was awarded a Nobel Prize in 1977.[1] The principle is simple. You take a sample that contains an unknown amount of a given hormone, and add to it a known amount of the same hormone labeled with a radioactive isotope and a small amount of an antibody to that hormone. Some of the labeled hormone and some of the unlabeled hormone in the sample will bind to the antibody. The binding is *competitive:* molecules of antibody have a limited number of binding sites, so the hormone in the sample competes with the labeled hormone for binding to these sites: less of the labeled hormone can bind when there is more unlabeled hormone in the sample. The bound fraction is then precipitated, usually by adding another antibody that recognizes the original antibody, forming a large complex that can be spun down into a pellet using a centrifuge. The amount of labeled hormone in the pellet is then measured by counting the radioactivity, and the values are compared with a standard curve generated by adding known concentrations of unlabeled hormone to aliquots of assay buffer, which are processed in the same way as the samples. The same principle is used in *enzyme-linked immunosorbent assays* (ELISAs), which differ from radioimmunoassays in using nonradioactive labeled hormones and a color reaction, rather than radioactivity, to measure binding.

Radioimmunoassays could be extremely sensitive and specific, were simple to perform, and could handle large numbers of samples. Antibodies could be raised to many types of molecules, and by 1975, twenty years after the first publication of the approach, more than 4,000 laboratories in hospitals, universities, and research institutes in the United States alone were running radioimmunoassays to measure many hormones and other biological molecules.

In Pittsburgh, Ernst Knobil and his colleagues developed a radioimmunoassay for LH. In rhesus monkeys, they used it to measure secretion in blood samples taken every ten minutes, and showed that LH was secreted in pulses that occurred at about once an hour.[2] Over the next few years, the full profile of LH and FSH secretion during the menstrual cycle of monkeys and women was revealed,[3] along with the pattern of secretion in males, and then the profiles in rats, guinea pigs, pigs, sheep, goats, cows, and horses. During the menstrual cycle of the rhesus monkey and of women, pulsatile secretion of LH and FSH is interrupted every month by a massive "surge" of these hormones, which, in the rhesus monkey, precedes ovulation by about 37 hours. A similar surge occurs in most female mammals, and it is generally an absolute prerequisite for ovulation. In all species, secretion is pulsatile at other stages of the ovarian cycle, and is pulsatile at all times in males.

The next step required understanding how LH pulses arise. Geoffrey Harris had hypothesized that secretion of hormones from the anterior pituitary depends on the secretion of releasing factors into the blood vessels that link the hypothalamus and pituitary. What exactly were these factors, if they existed at all (which some still disputed)? Rosalind Yalow shared her Nobel Prize with Roger Guillemin and with Andrew Schally, who, more or less simultaneously, discovered two of the releasing factors: thyrotropin-releasing hormone and GnRH.[4,5]

In the early 1960s, investigators had found that the hypothalamus contained a substance that stimulated LH secretion, and they provisionally named it "LH-releasing factor" (LRF). To get enough of this substance to enable it to be identified was a massive undertaking. From 160,000 hypothalami, Schally isolated 800 µg of LRF, just enough to determine its structure by the techniques then available. The deduced structure was confirmed by synthesizing the peptide, a process that also provided abundant supplies for further research. The synthesized peptide explained all of the biological activity of the hypothalamic extract—and it stimulated the secretion of not just LH but also FSH. This resolved the question of whether there was a separate releasing factor for FSH, and the early name LRF gave way to the present name, GnRH. GnRH has since been identified in many animal species from sea squirts to mammals. The amino acid sequence is well conserved, suggesting that it is evolutionally ancient, and in every species studied it has an important role in reproduction.[6]

Given that LH is secreted in pulses, it was natural to expect that these might be the result of pulses of GnRH, but this was not necessarily so. Many oscillatory phenomena in biology arise from constant stimulation; for example, the uterus of a pregnant female will contract rhythmically even when exposed to a constant concentration of oxytocin. To test the hypothesis that pulses of LH are the result of pulses of GnRH, it would be necessary to simultaneously measure the secretion of both. Measuring LH was not difficult, but GnRH is secreted at tiny amounts into the vessels that connect the hypothalamus to the pituitary. GnRH is also much smaller than LH, just ten amino acids, and measuring such small peptides by radioimmunoassay is much more difficult than measuring LH: blood samples must be processed to eliminate plasma matrix molecules that can interfere with immunoassays for small molecules. Moreover, any attempt to measure GnRH secretion would have to be made in a conscious and unstressed animal, since both anesthetics and stress stop pulsatile LH secretion.

Iain Clarke and James Cummins in Melbourne set out to do the seemingly impossible.[7] In sheep, they devised a way to collect blood samples from the *median eminence*, the specialized region at the base of the hypothalamus where small blood vessels collect the hypothalamic releasing factors. They implanted two needles into this region. Once the sheep had recovered, they used one needle to introduce a fine stylus to "nick" some of the vessels, and the other to collect the resulting blood, and they also collected blood samples from the jugular vein. The results were clear: every LH pulse was preceded by a GnRH pulse. The approach was refined by Alain Caraty in Nouzilly and Fred Karsch in Michigan, who revealed the relationship between LH and GnRH pulses in sheep in portal blood samples taken every 30 seconds. They also showed that the preovulatory LH surge is the consequence of a surge of GnRH.[8]

Knobil and colleagues showed just how very important the pulsatile pattern of GnRH is. In rhesus monkeys, lesions of the mediobasal hypothalamus abolish LH and FSH secretion by eliminating the supply of GnRH. Normal LH secretion could be restored by injections of GnRH, but their effectiveness depended more on the pattern of delivery than on the amount given. Continuous infusions failed to sustain LH secretion, which after an initial increase declined to undetectable levels. However, hourly injections, mimicking the physiological frequency of LH pulses, reproduced the physiological pattern of LH secretion. The explanation is that, when GnRH is infused continuously, the initial response of the pituitary is followed by a downregulation of LH and FSH secretion. As later summarized by Knobil, "the intermittency of the GnRH signal, within a relatively narrow window of frequencies, is an obligatory component of the neuroendocrine control system that governs normal gonadotropin secretion."[9] Knobil went on to note that "these fundamental physiologic observations in a nonhuman primate were transferred with remarkable rapidity to the clinical arena in the treatment of infertility that was attributable to hypothalamic dysfunction and in the suppression of inappropriate gonadotropin secretion (e.g., precocious puberty)."

The findings became important for treatments to enhance fertility, but also for new methods of contraception. The depression of LH secretion by continued administration of GnRH is so marked that GnRH agonists can be used as contraceptives for both males and females. Long-acting agonists are now used for this purpose in many domestic species, including dogs and cats.[10]

The pattern of LH secretion changes throughout life. Large amounts are secreted by the newborn infant, but then LH secretion drops to very low levels until it is reactivated at puberty. After puberty, the adult pattern of pulsatile secretion is established, followed by high levels of pulsatile secretion in late life as gonadal steroid levels decline. In the adult female, the pattern of LH secretion varies across the ovarian cycle in response to feedback from changing levels of estrogen and progesterone and other hormonal signals from the ovaries.

When GnRH binds to its receptors on gonadotrophs, intracellular calcium stores are mobilized, and this triggers exocytosis of vesicles. Every pulse of GnRH stimulates a pulse of LH, but FSH follows less consistently.[11] FSH *is* secreted in response to GnRH, but it is also secreted *constitutively*, at a rate that depends on the rate at which it is synthesized. The *pattern* of GnRH pulses makes a difference; low-frequency pulses favor FSH secretion, and high-frequency pulses favor LH secretion. Gonadal steroid and peptide hormones also modulate the synthesis of both FSH and LH. A variety of signals are involved, including three other hormones: *activin*, *inhibin*, and *follistatin*. Activin and inhibin are produced by the gonads (and also by pituitary cells) and have opposite effects: activin stimulates FSH production and inhibin inhibits it. Follistatin (and a variety of other messengers) is released from *folliculo-stellate cells* in the pituitary, curious cells of which we know little, but which appear to be an endocrine version of interneurons. These cells have long processes entwined among the endocrine cells, and they are electrically coupled. Follistatin binds activin, preventing it from stimulating the gonadotrophs, and so it is a *functional antagonist*, modulating the hormonal feedback from the gonads to the pituitary.

The *ovarian cycle* comprises the processes by which a set of follicles develops in the ovary to the point of ovulation, when the ova that they contain are released into the fallopian tube to become available for fertilization. Each follicle consists of an oocyte (an immature ovum) enclosed in a cluster of other cells, including granulosa cells that produce estrogen. At puberty, a woman will have several hundred thousand follicles in her ovaries, but normally only one will develop fully in each cycle. In a rat, the cycle occupies four or five days in three stages: *proestrus*, *estrus*, and *diestrus*. At the end of proestrus, a set of ovarian follicles begins to grow in response to pulsatile secretion of FSH from the pituitary. Soon, only a few follicles will continue to develop—the "leading" follicles secrete factors that suppress the development of less mature follicles. As the follicles develop, the granulosa cells proliferate and develop receptors for LH, and pulsatile secretion of LH stimulates them to produce more and more estrogen. The estrogen inhibits GnRH release, and hence the LH pulses become attenuated. Despite the attenuation of LH secretion, the increasing sensitivity of the follicles to LH means that estrogen levels continue to rise, reaching a peak in the afternoon of proestrus.

This escalating production of estrogen has many consequences. At the pituitary, it stimulates the production of LH and of GnRH receptors, and in the brain it activates neurons in the rostral hypothalamus that trigger a surge of GnRH secretion. This triggers the *preovulatory LH surge*, a surge that over a few hours depletes most of the large pituitary content of LH that has previously expanded in response to estrogen. The LH surge triggers ovulation in the afternoon of estrus: the follicles rupture, releasing ova into the fallopian tube, and their remnants form the *corpus luteum*. The corpus luteum produces a surge of progesterone coincident with the LH surge, and throughout diestrus it continues to produce progesterone until it is destroyed in a process called *luteolysis*, the mechanisms of which vary considerably between species. The high level of progesterone also acts at both the pituitary and the hypothalamus, suppressing the secretion of GnRH, LH, and FSH.

While continuing high levels of estrogen are needed for an LH surge, the mechanisms that drive the surge are present only in females—the hypothalamus is sexually dimorphic. The exact timing of the surge depends on other signals. The cat is a *reflex ovulator*: the LH surge in cats is triggered by coitus. Rats are *spontaneous ovulators*: the LH surge occurs at a fixed time of day, and its timing is governed by the suprachiasmatic nucleus of the hypothalamus. Many species are *seasonal breeders*: in them, a prolonged period of infertility is terminated by environmental cues, such as day length and ambient temperature, to ensure that young are born at a time when food is abundant. In many of these species, pheromonal signals from urine, feces, vaginal secretions, saliva, and specialized scent glands help to synchronize reproductive cycles. This ensures that the young are all born at about the same time; a greater proportion of them will survive when their predators are saturated with prey.

The actions of estrogen and progesterone in the brain are not confined to the control of GnRH secretion. The term *estrus*, from the Greek word for a gadfly, was coined in 1900 to describe the "special period of sexual desire of the female."[12] The human is unusual in being potentially willing to engage in sexual activity at any time of the cycle: in most mammals, the female is receptive to males only at estrus, just after ovulation, and this receptivity is governed by the actions of estrogen and progesterone on the rostral hypothalamus.

This is a *very* simplified account. The details of the ovarian cycle differ between species, but all involve reciprocal interactions between the brain, the pituitary, and the ovary, each of which has a pattern-generating machinery of its own, involving many cell types and many signals. When three complex systems interact reciprocally to generate a cycle, it makes little sense to call any one a master and any other a slave. The gonads control the brain and pituitary as much as the brain controls the others.

Endocrine cells are not passive; they form an organized community using multiple messaging systems to generate complex patterns of secretion, and the regulation of synthesis is as important as the regulation of secretion. The intrinsic properties of the gonadotrophs amplify and smooth the pulsatile signals from GnRH neurons to generate large and orderly LH pulses, but their signal-generating abilities really come into their own during the preovulatory LH surge.

The LH surge is triggered by a surge of GnRH, but it is far from a passive response. The extraordinary process that underlies this surge was unmasked by George Fink, who called it the "self-priming" actions of GnRH.[13] Although it is the actions of steroids on the brain that ultimately result in the LH surge, steroid actions at the pituitary are also important. By increasing the synthesis of LH, estrogen increases the amount of LH that can be secreted, and by increasing the synthesis of GnRH receptors it enhances the sensitivity of gonadotrophs to GnRH. Also, and most importantly, gonadotrophs that have been exposed to high levels of estrogen for long enough *change how they respond to GnRH* (figure 6.1).

To understand this, we might ask a simple question, a question that can be asked both about release of neurotransmitters at synapses and about hormone secretion from endocrine cells. In both cases, secretion results from *exocytosis* of vesicles, either the small synaptic vesicles that contain neurotransmitter or the large dense-core vesicles that contain peptides. In both cases, exocytosis is triggered by an increase in intracellular calcium. At synapses, exocytosis is the result of calcium entry through channels opened by depolarization. In endocrine cells, exocytosis can result from chemical signals that activate signaling pathways leading to mobilization of intracellular calcium stores. In both cases we can ask why only *some* vesicles are released. At synapses, the answer is that, at any given time, only a few synaptic vesicles are available for release. Exocytosis of synaptic vesicles can occur only at specialized sites, close to clusters of calcium channels. To be available for release, a vesicle must be "docked" at one of those sites, where spikes cause a local rise in calcium and where particular proteins enable the synaptic vesicle to fuse with the cell membrane. As one vesicle is released from one of these sites, another takes its place, awaiting the next signal. In endocrine cells, however, calcium signals from intracellular stores spread throughout the cell and are relatively prolonged. The secretory response must be carefully rationed, and when some vesicles are secreted others must be moved around from deep within the cells.

It's complicated: you might expect it to be like a tube of marbles in which, when the marble at the bottom is released, a new marble enters at the top, but it is not. Newly synthesized vesicles go first to the plasma membrane to be part of a "rapid-release" pool. If they are not released, they are shuttled into a "reserve pool" from which they can reenter the rapid-release pool if it becomes depleted. If they are still not called upon, they enter a

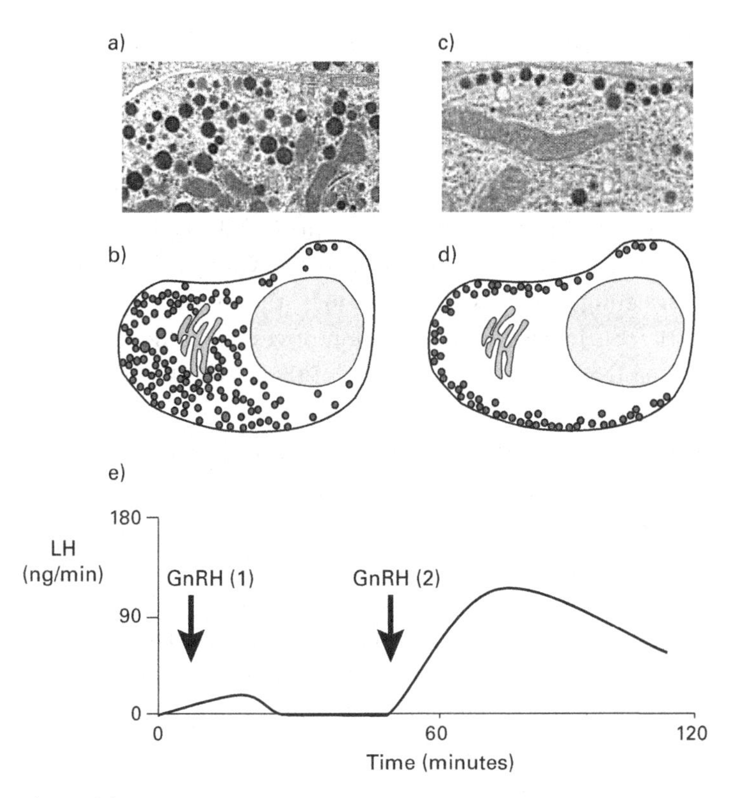

Figure 6.1
Self-priming. In gonadotrophs, LH and FSH are packaged in large dense-core vesicles. (a) shows an electron microscope image of these vesicles, which appear as small, dense, spherical objects that are scattered throughout the cell.[14] (b) shows a reconstructed cross section of a gonadotroph, showing the nucleus (N) and the endoplasmic reticulum. In a proestrus rat, exposure of the pituitary to GnRH causes an increase in intracellular calcium that comes from the stores in the endoplasmic reticulum. This causes some LH to be secreted, but it also causes the remaining vesicles to be trafficked to sites close to the cell membrane. (c) shows the appearance of such a gonadotroph, and (d) shows a schematic reconstruction of the whole gonadotroph. Now, more of the vesicles are available to be released, so when a second GnRH challenge is given, much more LH is secreted. The graph (e) shows the resulting secretion. Redrawn from Scullion et al.[15]

"nonreleasable pool" where they will be dismantled. This flow is regulated by a "scaffold" of actin filaments. Cells are not mere bags of stuff, they have an internal *cytoskeleton* of contractile filaments that organize the internal compartments and enable cells to move.

The pituitary of an ovariectomized rat responds consistently to repeated challenges with GnRH, secreting similar amounts of LH each time. However, in a pituitary that has been exposed to estrogen, GnRH pulses result in progressively escalating responses. GnRH "primes" the response of gonadotrophs to GnRH—this involves a translocation of vesicles to docking sites at the plasma membrane, which makes more of them available for release (figure 6.1). As a consequence, in some species an LH surge can be evoked without *any* GnRH surge. In rhesus monkeys, after hypothalamic lesions that remove all GnRH, ovarian cycles can be restored by hourly pulses of GnRH, and *no change in this pattern* is needed to produce the monthly surge of LH and FSH that triggers ovulation. The feedback actions of sex steroids, by modulating how the pituitary responds to GnRH, are enough.

Thus secretion from endocrine cells depends on the signals that they receive, on their sensitivity to those signals, on paracrine interactions among the endocrine cells and with neighboring cell types, on the number of vesicles available for secretion, and on where those vesicles are located within the cell. The rates of synthesis of hormones and receptors, the intracellular disposition of vesicles, and intercellular organization and communication are all dynamically regulated.

Any notion that endocrine cells are merely slaves to the brain's bidding is mistaken. The gonadotrophs are not exceptional in being organized as a community, or in being active partners in pattern generation.

Growth hormone, for example, is secreted in large pulses, the amplitude and frequency of which determine the rate of body growth. In species such as rats where there is a large disparity between the size of males and females, the pattern of secretion is sexually dimorphic. In male rats, a large pulse is secreted every three hours or so, and the importance of this pattern was revealed by Iain Robinson and his colleagues at the National Institute for Medical Research at Mill Hill, London. They studied rats deficient in growth hormone[16] and measured their growth rates in response to injections of growth hormone. They showed that growth rates typical of normal male rats could be achieved in either male or female rats if growth hormone was injected at similar intervals to those seen in normal male rats, but not if the same amount was delivered by constant infusion or given as larger, less frequent injections.

Robinson went on to study how the pulses of growth hormone are generated. He found that male rats released more growth hormone in response to injections of GHRH than female rats, but whereas the responses in female rats were very consistent, the responses in male rats were erratic. When injections were given every 90 minutes, male rats responded strongly only to every other injection. Robinson concluded that another factor was intermittently suppressing the response of the somatotrophs to GHRH.

We now know, thanks in large part to Robinson's studies, that pulses of growth hormone secretion arise from an interplay at the pituitary between GHRH, which stimulates growth hormone secretion, and a second hypothalamic peptide, somatostatin, which inhibits it. When GHRH triggers a pulse of growth hormone secretion, that pulse triggers the secretion of another hormone, *insulin-like growth factor-1,* from the liver, which enters the brain to act on the somatostatin neurons. The somatostatin that these neurons then release acts at the pituitary to prevent the somatotrophs from responding to GHRH. Only when the somatostatin signal has died away can GHRH trigger another pulse of growth hormone. These periventricular somatostatin neurons are one of the sexually dimorphic populations of the hypothalamus.

An even stronger role of peripheral endocrine mechanisms is apparent in the case of the "stress hormone" ACTH. ACTH is released from the corticotroph cells of the pituitary in pulses that stimulate pulsatile production of glucocorticoids that act back on the corticotrophs to inhibit them. As revealed by an elegant mathematical model, this partnership of endocrine cells can, by the exchange of hormonal signals, generate pulsatile secretion without any involvement of the hypothalamus.[17] In this case, the hypothalamic releasing factors seem to modulate an autonomous endocrine pulse generator, rather than dictate the pattern of endocrine secretion.

Thus endocrine cells have a repertoire of properties that are not normally associated with neurons, but which nevertheless have sophisticated computational implications. These properties are also present in some hypothalamic neurons. Generally, the temporal patterning of peptide signals in the periphery is an important determinant of their biological efficacy. Given that peptides are commonly used signals within the brain, it seems likely that, there too, the temporal pattern of peptide secretion is important.

7 Dendritic Secretion and Priming

Dans les champs de l'observation le hasard ne favorise que les esprits préparés. (In the fields of observation chance favors only the prepared mind.)

—Louis Pasteur (1822–1895), lecture, University of Lille, December 7, 1854

In the thirty years after Wakerley and Lincoln's description of the milk-ejection reflex, the question of the origin of milk-ejection bursts inspired a profusion of hypotheses. The most obvious notion was that the bursts reflected some patterning of the suckling input. In a mother with only one suckling young, it is conceivable that, with time since the last delivery of milk, the sucking might become more urgent, triggering the next delivery. A mother rat usually has about ten pups suckling: sometimes one becomes agitated and sucks particularly vigorously, but usually the others do not join in. Nevertheless, the signal from the litter *might* show periodic variations that are hard to detect; recordings of activity in the mammary nerves indeed showed that the signals from the nipples were very erratic.[1,2] Two observations destroyed this hypothesis. First, injecting oxytocin to produce milk letdown had no effect on the ongoing pattern of milk ejections, dispelling the idea that it is the time since the last delivery of milk that determines the timing of bursts. Second, in the absence of any suckling pups, a normal pattern of milk ejections could be established by mild electrical stimulation of the nipples, delivered randomly.[3]

Another notion was that the suckling input was integrated in a "pattern generator" somewhere in the brain that sent its output to the oxytocin cells. The main rationale for this proposal was that the supraoptic nucleus on one side of the brain is separated from the other supraoptic nucleus by the third ventricle. Accordingly, perhaps some midline structure projected to both nuclei to engage them simultaneously in the reflex. This notion was harder to dispel, because it was (and still is) unclear which brain areas are involved in transferring suckling information to the hypothalamus. Experiments using lesions of different brain regions excluded some areas; areas rostral to the hypothalamus were clearly not needed, for example, but areas of the caudal brainstem that receive spinal inputs could not be lesioned without grave effects on respiration or blood pressure. One candidate area was the ventral medulla, which contains neurons that express c-*fos* in response to suckling. C-*fos* is a gene that is widely used as a marker of functional activation of neurons: it is expressed in many neurons for a short while after they have been activated, and encodes a transcription factor that can stimulate the expression of many other genes.[4] C-*fos* is one of several *immediate-early genes* that link neuronal activity to the activation of synthesis, to replace what has been secreted. The protein product of c-*fos* appears in the nuclei of neurons within about a half hour of their activation, and remains there for another hour or so. This product can be detected by immunocytochemistry—by exposing brain sections to an antibody to the c-*fos* protein and then visualizing the antibody with a color reaction. It's a powerful technique that allows the impact of a particular stimulus on neurons throughout the brain to be visualized.

Neurons of the ventrolateral medulla that use noradrenaline as a neurotransmitter project to the supraoptic and paraventricular nuclei, but mainly to vasopressin neurons, not oxytocin neurons. However, just medial to the noradrenergic neurons, Françoise Moos and her colleagues found some neurons that projected to the supraoptic nuclei.[5] Their axons remained on each side of the brain until they reached the level of the supraoptic nucleus, where they gave rise to branches, some of which entered the optic chiasm to cross the midline below the third ventricle and reach the opposite supraoptic nucleus. When lesions were made in the middle of the optic chiasm, some rats still displayed milk-ejection bursts, but the bursts no longer occurred at the same time in the two nuclei. This result supported the notion that these neurons carried the suckling input. However, it did not show that their activity was organized into a bursting pattern, and if these axons could cross the midline below the third ventricle to link the supraoptic nuclei, perhaps other axons did too.

The main reason for thinking that the bursts must originate close to the magnocellular neurons was that injecting tiny amounts of oxytocin into just one supraoptic nucleus could trigger milk-ejection bursts in both that nucleus and the contralateral nucleus.[6] Was the oxytocin acting on the oxytocin cells themselves, or on some nearby neurons that generated the bursts and projected to all four nuclei, perhaps by axons that crossed the midline?

When the distribution of oxytocin receptors in the brain was first described, receptors were found in many places, but, frustratingly, apparently not in the oxytocin cells.[7] Nevertheless, oxytocin cells seemed to have functional receptors: oxytocin could trigger mobilization of calcium stores in isolated oxytocin cells, and oxytocin could trigger oxytocin release from explants of the supraoptic nucleus in vitro. When it became possible to visualize the mRNA for the oxytocin receptor, it was clear that the gene *was* expressed in the oxytocin cells.[8] So why was it so hard to see the receptors? The answer is that most of the receptors were internalized—inside the cells—and invisible to the techniques used to locate them. Many peptide receptors are internalized after ligand binding and return only slowly to the cell surface, so perhaps supraoptic neurons are bathed in a constant excess of oxytocin and the receptors only transiently appear on the cell surface. Knowing that oxytocin could cause oxytocin release, Marie-José Freund-Mercier and her colleagues injected rats with an oxytocin antagonist into the brain to try to reduce this release: their bold experiment worked; in antagonist-treated rats they found abundant oxytocin receptors on the dendrites.[9]

Thus oxytocin *does* act on oxytocin cells, and can mobilize intracellular calcium stores and trigger oxytocin release. But oxytocin cells have just one long axon that does not branch inside the supraoptic nucleus—there are some oxytocin-containing synapses there,[10] but not enough to get excited about—and the dogma was that without axon terminals there could be no physiologically regulated release.

A different idea was that the oxytocin release measured from explants of the supraoptic nucleus came from just outside the nucleus, where short branches of some of the axons end on other neurons. Richard Dyball and I found some neurons outside the nucleus that received inputs from magnocellular neurons, but none responded much to suckling, and certainly not with anything like a milk-ejection burst.[11]

If oxytocin cells talked to each other, either directly or indirectly via interneurons, this should not have been hard to show. When I began studying the supraoptic nucleus the dogma was that there *were* recurrent pathways, but they were inhibitory, not excitatory. Whenever a stimulus was applied to the neural stalk to antidromically evoke a spike in a supraoptic neuron, it fell silent for about 50 milliseconds, and this had been interpreted as evidence for recurrent inhibition—it was assumed that the activity evoked in oxytocin cells resulted in activation of interneurons that projected back to inhibit the oxytocin cells. I thought it more likely that no interneurons were involved, and that the spikes triggered an intrinsic hyperpolarization.[12] Soon, intracellular recordings revealed that spikes were indeed followed by a hyperpolarizing afterpotential, and the notion of recurrent inhibition was quietly forgotten.[13]

I pressed on to look for evidence of cross talk. The perfect experiment would be to record from one supraoptic neuron while applying a stimulus that would activate every supraoptic neuron *except* the cell that I was recording from, and a trick let me do something like that. When a stimulus is applied to the neural stalk, an antidromic spike is evoked in the axon of every supraoptic neuron. But not all of these spikes will reach the cell bodies: if an antidromic spike traveling up an axon meets a spontaneous spike traveling down, both will be snuffed out—the process is called *collision*. I set my stimulator to generate stimuli triggered by the spontaneous spikes from a supraoptic neuron. These stimuli had no direct effect on the cell I was recording, but activated every *other* supraoptic neuron. The experiments revealed that there *was* cross talk:[14] after a few minutes of stimulation, the oxytocin neurons started to show burst-like activity. It seemed that activating the neurons had a mixture of effects that could not easily be explained by conventional synaptic interactions.

Could something be released from the oxytocin cells at sites other than synapses? The dendrites were full of oxytocin; it was assumed this was a store waiting to be sent down the axons, but perhaps it could be released. John Morris and his student David Pow found a way to test this.[15] Tannic acid binds to extracellular proteins but does not enter cells. In tissues exposed to tannic acid, the dense cores of protein-rich vesicles that are in the process of being released are firmly "fixed" and can be visualized by electron microscopy. By using tannic acid, Morris and Pow could find exactly where vesicles were being released, and at what rate. In brain slices exposed to tannic acid, they found that strong depolarization caused vesicles to be released from the soma and dendrites of magnocellular neurons, and they made the bold claim that "exocytosis from dendrites could well account for a large part of the release of oxytocin and vasopressin into the hypothalamus and cerebrospinal fluid."

Mike Ludwig and I set out to see if dendritic release could be evoked by spike activity in intact rats. He had trained in Leipzig with Rainer Landgraf, who had pioneered the use of microdialysis to measure peptide release in the brain, and he refined the technique to combine it with electrophysiological recordings. Microdialysis uses a tiny probe, the tip of which is a semipermeable membrane. The probe is inserted into a discrete brain region, and artificial cerebrospinal fluid is pumped slowly through it; if peptides are present, some will cross the membrane and can be measured in fractions of the perfusate by a sensitive radioimmunoassay. We set out to see if evoking spike activity would stimulate dendritic oxytocin or vasopressin release. The results were dispiriting: even intense stimulation that released massive amounts of oxytocin and vasopressin into the blood had no detectable effect on release from the soma and dendrites.[16]

I wasn't too surprised. If bursting activity was simply the result of a positive feedback signal that resulted from spike activity, then any stimulus that intensely activated the oxytocin cells should produce bursts. This was not the case; many stimuli could excite the neurons, yet bursts were never seen except in response to suckling or during parturition. So, was something about the neurons different in lactation?

Dionysia Theodosis in Bordeaux[17] and Glenn Hatton in Michigan[18] found that something was indeed happening in the supraoptic nucleus at the end of pregnancy. In a male rat or a virgin rat, the dendrites of magnocellular neurons do not lie directly next to each other, but are separated by thin, filamentous processes of glial cells, like insulating sheaths. At the end of pregnancy, these processes retract, leaving the dendrites directly apposed in bundles of about five to ten dendrites. Only the oxytocin dendrites were affected, and Theodosis showed that oxytocin itself could cause this reorganization. Moreover, when the processes were retracted, "double synapses" appeared—single axons penetrating the nucleus were seen to make contacts with multiple dendrites. So two possibilities became apparent: perhaps the retraction of glial processes allowed oxytocin cells to interact directly, or perhaps it meant that certain inputs were shared among oxytocin cells; either possibility might help to explain how oxytocin cells came to be synchronized.

But Theodosis found that the shared synapses were mainly inhibitory, not excitatory.[19] Were these even relevant to the milk-ejection reflex? Axons that enter the supraoptic nucleus typically make contact not with a single neuron but with many; perhaps it was just easier to see multiple contacts when the glial cell processes were retracted.

In Michigan, Glenn Hatton thought that direct contacts between the dendrites might permit direct electrical interactions.[20] Gap junctions are a mechanism of electrically coupling neurons that is common in insects and other invertebrates, and it underlies synchronization of spike activity in some other areas of the brain. Hatton looked for gap junctions in the supraoptic nucleus by filling neurons with a dye that can cross them. In most cases, the dye was found only in the cell that he had injected, but sometimes it spread to two or three other neurons. This seemed too limited a connectivity to account for synchronization of all the neurons. Moreover, the dye coupling seemed to be as common between vasopressin neurons as between oxytocin cells, but vasopressin neurons always fire *asynchronously*.

Then, in a fine example of Popperian science, Theodosis devised a "killer experiment" to test her hypothesis that the glial reorganization was critical for the milk-ejection reflex. She had shown that this reorganization depended upon neural adhesion molecules, and could be blocked by endoneuraminidase, an enzyme that inactivates the adhesion molecules. She now found that, after endoneuraminidase injections, both parturition and the suckling-induced milk-ejection reflex were unaffected. Theodosis and her colleagues accepted their own refutation of their bold hypothesis, concluding that "neuro-glial remodelling is not essential to parturition and lactation."[21]

One obstacle to understanding the reflex was that oxytocin neurons in hypothalamic slices in vitro didn't seem to want to show anything like milk-ejection bursts. Glenn Hatton with Yu-Feng Wang had tried harder than anyone, and eventually persuaded some neurons to display realistic-looking bursts, but only by employing a combination of pharmacological challenges and exotic conditions that gave little clear indication of how bursts might be generated in a living animal.[22] Jean-Marc Israel and his colleagues in Bordeaux set out on another tack.[23] They cultured explants of the hypothalamus from newborn rats, and found that oxytocin cells in these cultures generated synchronized bursts that sometimes looked quite like milk-ejection bursts, and which could be facilitated by adding oxytocin to the cultures. These bursts seemed to depend on interaction between the oxytocin cells and other glutamate neurons in the culture. Their experiments were beautiful, but also depressing: they didn't *explain* the origin of bursts, they deferred that question back to some unknown postulated population of glutamate neurons with which the oxytocin cells interact.

But neither Françoise Moos nor I were convinced. The oxytocin cells in cultures showed correlated spiking activity independently of any bursts, and this was something we had never seen in living animals. Nor did the bursts seem as intense as milk-ejection bursts, or have quite the same profile. In newborn animals, oxytocin cells have complex branching processes that contact many other neurons; over the first few days of life, most of these branches disappear, leaving just one long axon that projects to the posterior pituitary and one or two unbranching dendrites.[24] The oxytocin cells don't only release oxytocin, they can also release glutamate from their axonal terminals. It seemed possible that the cultures preserved the extensive connectivity of newborn animals, and that the sparseness of neurons in a monolayer culture meant that a greater proportion of their inputs would be shared.

Moos and I set ourselves aside to review the wreckage of discarded and refuted theories about the milk-ejection reflex. She had been studying the reflex for longer and more intensely than anyone. Dendritic oxytocin release had to be essential: oxytocin could be released from dendrites; it was released by suckling even before any milk-ejection bursts; it acted directly on the oxytocin cells to cause more oxytocin to be released; and it had some excitatory effects. There was *some* kind of activity-dependent positive feedback, probably involving oxytocin, but this was present only intermittently. Injecting oxytocin into the brain, or even into just one supraoptic nucleus could trigger a sequence of intense milk-ejection bursts only in a suckled, lactating rat. Hideo Negoro and his colleagues had applied my trick of stimulating the neural stalk in a pattern triggered by the spontaneous spikes from a supraoptic neuron to show that this could similarly trigger a sequence of intense bursts in suckled, lactating rats, a much stronger effect than I had seen in nonlactating rats.[25]

We assembled a list of other features of the reflex that we thought were well established but paradoxical—features that had caused the many previous ideas to flounder.

1. The oxytocin cells, even in a lactating rat that is being suckled, are *not* strongly synchronized. Moos had made simultaneous recordings of pairs of oxytocin cells; sometimes of adjacent neurons in one nucleus, sometimes of an oxytocin cell in one supraoptic nucleus and another on the other side of the brain. The bursts never began as soon as suckling started—there was always a latency of many minutes, and sometimes a burst occurred in the few minutes after the pups had been removed from the nipples—it thus seemed that suckling had a delayed but persistent effect on the oxytocin cells.
2. Bursts arose at about the same time in all neurons—but there could be a difference of a few hundred milliseconds.[26] This varied from burst to burst—one neuron was never consistently the leader and the other the follower. Bursts arose at the same time in all nuclei. It seemed that bursts could be triggered by the activity of just a few oxytocin neurons that then spread to all of the others.
3. The spikes were never simultaneous, even during a burst, and were no more closely correlated between adjacent neurons than between neurons on opposite sides of the brain. After a burst, when activity first returned to normal, the activity of pairs of cells was always uncorrelated. However, shortly before the next burst, the spike activity became irregular: Moos, who had listened to many hundreds of these bursts, could hear this change, and so could predict when a burst would come. These fluctuations *were* correlated between neurons, but only weakly: it seemed that the oxytocin neurons were not always in communication with each other, but only at about the time of the bursts.
4. Increasing the activity of oxytocin cells with any of many different stimuli typically stopped the reflex, at least for a while, but, paradoxically, inhibiting the neurons sometimes triggered a burst.[27]
5. After a burst, the neurons fell silent for many seconds, and sometimes one fell silent after a reflex milk ejection without discharging any burst at all. This was a curious observation that both Moos and I had made, and it suggested that, when the oxytocin cells fell silent after a burst, it was probably *not* simply activity-dependent inhibition.

Instead of treating these observations as anomalies to be disregarded, we took the opposite position, that these anomalies had to be explained. The idea that we came to was inspired by the "self-priming" actions of GnRH discovered by George Fink.[28] In gonadotroph cells of an estrogen-primed rat, a pulse of GnRH can cause LH vesicles to be relocated to near the plasma membrane, where they are available for release in response to a subsequent challenge with GnRH. We wondered if something similar might happen in oxytocin cells, and Mike Ludwig and I set about to test this.

When oxytocin acts on oxytocin cells it induces a mobilization of intracellular calcium stores, as GnRH does in gonadotrophs. Ludwig and I tried again to measure oxytocin secretion from dendrites, and now we looked at the effects of mobilizing intracellular calcium. We used *thapsigargin*, which blocks the pump that cells use to fill their intracellular stores with calcium; when cells are exposed to thapsigargin, calcium flows out from these stores, which then can't be refilled. When we exposed oxytocin cells to thapsigargin, we saw some release of oxytocin, and when we then stimulated the neural stalk to evoke spike activity, we saw a very large release.[29] Thapsigargin had caused vesicles to be relocated into a pool from which they could be released by spike activity, and we called this phenomenon "priming" in deference to the term that Fink had used. We went on to show that oxytocin itself and some other peptides could also both cause immediate oxytocin release and prime activity-dependent oxytocin release[30,31] (figure 7.1).

Some peptides can trigger secretion from dendrites while *inhibiting* electrical activity. The clearest example of this is α-MSH.[32] α-MSH is released into the supraoptic and paraventricular nuclei from axons that project from the arcuate nucleus, and acts at MC4 receptors to mobilize intracellular calcium stores. This not only triggers oxytocin release from dendrites and primes oxytocin release, it also increases the production of *endocannabinoids* by the oxytocin cells. In the supraoptic nucleus, endocannabinoids inhibit glutamate release from afferent nerve endings, thereby reducing the spike activity of the oxytocin cells. Thus α-MSH suppresses oxytocin secretion from the pituitary while stimulating oxytocin release in the hypothalamus. By combining properties of neurons and endocrine cells, oxytocin cells *independently* regulate what they release centrally and peripherally.

So our explanation of the reflex was taking shape. When a lactating rat is suckled, that input releases a peptide into the magnocellular nuclei which causes some oxytocin to be released from the dendrites. This input signal also primes some of the dendritic vesicles, making them available for activity-dependent release. If the oxytocin cells are very active, primed vesicles will be released. The calcium mobilization also generates the production of

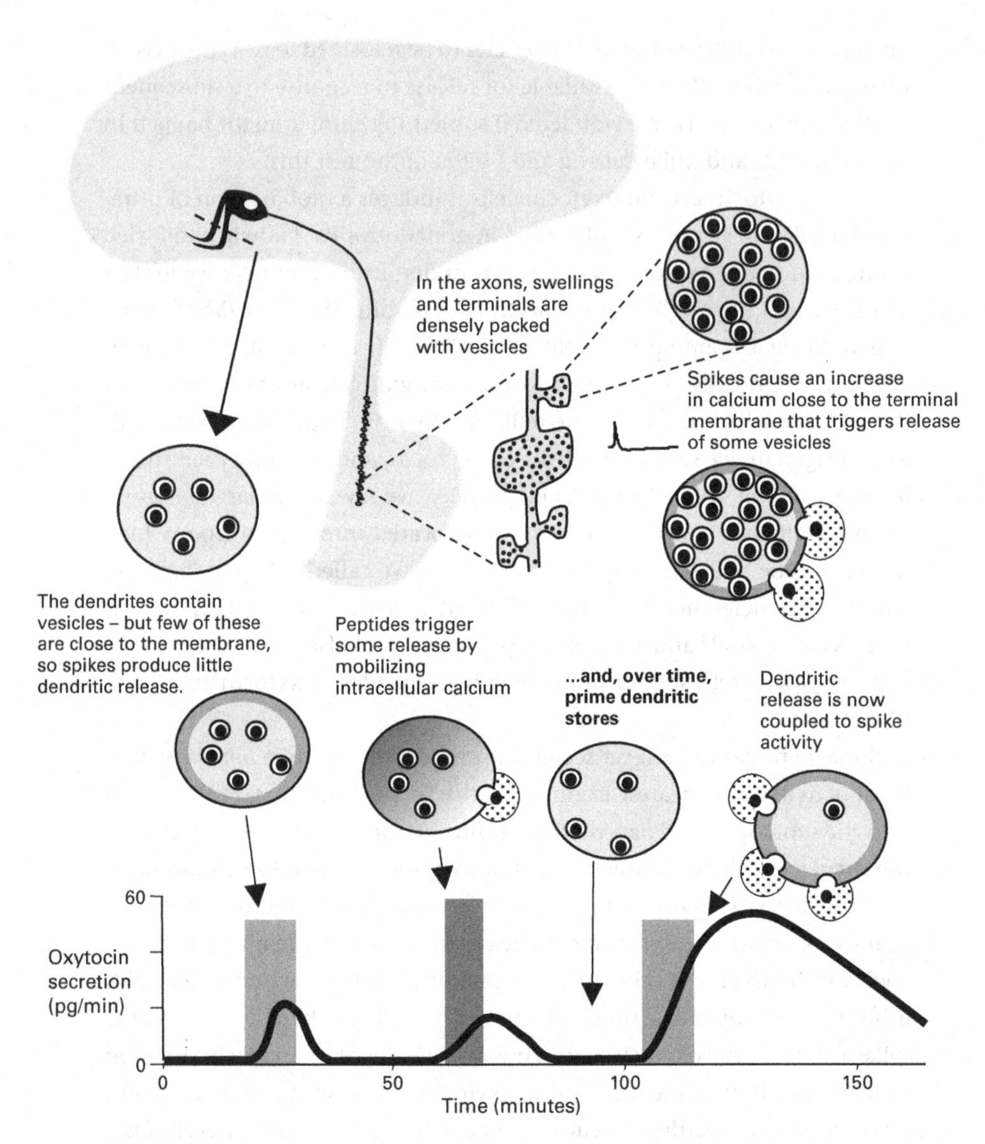

Figure 7.1
Priming in oxytocin neurons.

endocannabinoids, which suppress excitatory afferent inputs, keeping the neurons relatively quiet. As suckling proceeds, the store of releasable vesicles builds up; a few of these will be released erratically and will excite neighboring cells. Increasingly, the oxytocin cells "listen" more and more to each other and less and less to their afferent inputs, and their activity becomes more erratic but also increasingly coordinated. A critical point is reached when the pool of primed vesicles is large enough: a fluctuation in release will then trigger a burst of spikes in one cell that is propagated through all of the neurons, emptying their releasable pools.

To test the plausibility of this explanation, Moos and I, with mathematician colleagues in Rome and Cambridge, constructed a computational model.[33] We gave each of a population of model oxytocin cells a hyperpolarizing afterpotential and other intrinsic membrane properties known from intracellular recording studies. We gave each cell two "dendrites," randomly linked in bundles of about eight dendrites from different neurons. We gave each dendrite a "store" of vesicles that could be trafficked in response to suckling into a readily releasable pool. We proposed that release would be governed nonlinearly by spike activity in the same way that John Bicknell and I had shown that release from the nerve terminals of oxytocin cells was governed. We gave each neuron its own random "synaptic input" that could be attenuated by "endocannabinoids" whenever a burst was generated. When one oxytocin vesicle was released it had little effect, but if vesicles were released at about the same time from several dendrites in a bundle this would produce a strong depolarization that would cause further release.

The model met our expectations: to both of us the spiking activity of the model neurons was indistinguishable from the spiking activity we had recorded in real neurons. In response to a simulated "suckling input" that primed the releasable stores of oxytocin in the dendrites, after a delay of a few minutes the population of model cells began to show intermittent intense bursts of spikes that were identical to those seen in real cells. The model could also reproduce all of our list of paradoxical observations, and it also produced one unexpected feature. Because bursts produced both excitatory oxytocin release and inhibitory endocannabinoid production, in any given model neuron about one burst in three was preceded by a brief inhibition. We went back to our recordings of milk-ejection bursts and found, to our relief, exactly the same phenomenon in those.

So, for me, the job was done: we had an explanation that held up. The model is not simple; it wasn't explicitly designed to burst, but to display properties that we knew the oxytocin cells possessed to see if they might amount to an explanation of bursting. It's not an easy model to explain, because it depends on randomness and heterogeneity. The neurons *must* receive a random input: without random fluctuations of firing rate it doesn't work. The dendritic connections must be random: if cells are connected only to their nearest neighbors, the model gives not simultaneous bursts but a traveling wave of bursts. There must be *enough* cells in the network, the interactions between them must be weak, and the connectivity must be sparse. Because of these features, we described bursting in this model as *emergent behavior*, behavior that comes about naturally when a certain level of random complexity in the network is exceeded.

The model simplifies many things and leaves out many others. There aren't any glial cells in the model, though implicitly their influence is there in the pattern of sparse dendritic connectivity. We have only one nucleus, not four. How bursts are propagated among the nuclei is not clear. From the model we know that if only a few dendrites spread from one nucleus to another, this could be enough, but it is possible that other mechanisms are involved. Some parvocellular oxytocin cells in the paraventricular nucleus project to the supraoptic nucleus,[34] and perhaps these are involved. It is still unclear what is released from the suckling input to trigger priming in the first place. It may not be just oxytocin that excites other oxytocin cells—perhaps glutamate released from these neurons also has a role.

We didn't have all the answers. But this was a good place to stop, and a good place to start, because with the discovery of dendritic secretion and priming we had found phenomena that seemed to have a potential to help us understand how peptides could have long-lasting effects on complex behaviors. First, dendritic peptide secretion was functionally important and likely to be a common feature of neurons. Second, oxytocin could, by its ability to prime dendritic release, change the way that oxytocin cells interacted with each other—in effect, it could rewire neuronal networks. Third, oxytocin could trigger its own release, so its effects could be self-sustaining and long-lasting. Fourth, oxytocin cells could regulate release from their dendrites independently of release from their axon terminals.

8 The GnRH Neuron

Say not the struggle nought availeth,
The labor and the wounds are vain,
The enemy faints not, nor faileth,
And as things have been they remain.
If hopes were dupes, fears may be liars;
It may be, in yon smoke concealed,
Your comrades chase e'en now the fliers,
And, but for you, possess the field.
—Arthur Hugh Clough (1819–1861), "Say Not the Struggle Nought Availeth"[1]

When talking about oxytocin neurons, it seems possible to convey an understanding of much of what they do, of how and why they do it. That understanding did not come easily: anomalies, confusions, and misconceptions were gradually resolved; many gaps in our knowledge were first revealed and then filled in by many laboratories. My account has hidden most of this process of excavation, filtering, and refinement, and in so doing it misrepresents the nature of science and discovery.

We have a stable understanding of how oxytocin cells generate the milk-ejection reflex. There are uncertainties, but the shape of our understanding is, for the moment, clear (though it might yet be wrong). For GnRH neurons, our theories are still churning. This chapter is longer than most and more confusing, because here I am trying to convey what our science is like, how we grapple with inconsistencies and uncertainty. We have not answered the fundamental questions about how LH pulses and the LH surge are generated. Perhaps we have broken the problems; we know better what we don't know and need to know. By the time you read this, perhaps some of those gaps will have been filled.

A complete understanding of any physiological system must embrace many things. We need to know what it does that is important: how it contributes to the fitness of the organism of which it is a part. We also need to know the mechanisms by which it does whatever it does. Both of these require us to know how the system is determined by genetic codes, how it has evolved, and how it develops in early life.

Development is a rule-governed process. Cells born in a particular place at a particular time express just some of all the genes in an organism. That set of expressed genes confers properties that cause cells to develop and migrate in a particular way. As they migrate, they encounter other cells, and the signals that they receive change them further, suppressing some genes and engaging others—and the cells that they encounter are themselves changed by that encounter. By the end of development, we have a brain of immense complexity. If we knew all the rules, perhaps we could build something very like a brain in silicon. We might yet be able to build a brain in this way before we understood anything about its function.

In all mammals, GnRH neurons are essential for reproduction. You might imagine that, being so important, there would be many of them, but in the rat, there are only about 800. In humans, there might be about 10,000; this is how many there are in the fetus, though probably not all of them survive into adulthood.[2] Of the 10,000 GnRH neurons in the human fetus, about 2,000 are in the hypothalamus and are involved in regulating the pituitary. Other, "extrahypothalamic" GnRH neurons are in the olfactory bulbs, cerebral cortex, hippocampus, and a few other places, but what they do is unknown.

The GnRH neurons are extraordinary. They start life not in the brain, but in the nose. In mice, they are born in the nasal placode on the tenth day of embryonic life, together with olfactory sensory neurons. The sensory cells stay in the nose and form the *vomeronasal organ*, which is responsible for detecting pheromones—chemical signals that are important in many mammals for social and sexual behaviors. These sensory cells send axons into the olfactory bulbs, at the front of the brain, and the GnRH neurons migrate along them. The axons end at the outermost layer of the bulbs, but the GnRH neurons journey on through the developing forebrain. In rodents, about 800 end up scattered between the olfactory bulbs and the caudal hypothalamus. In vertebrate species from fish to humans, the hypothalamic GnRH neurons project their axons to the median eminence, where they end next to the portal blood vessels that transport releasing factors to the anterior pituitary.

At the National Institutes of Health in Bethesda, Susan Wray has long studied this migration, and in elegant and sophisticated experiments she has revealed many of the mechanisms involved.[3] The migration is guided by "chemotactic" cues from the cells that GnRH neurons encounter along the way, and defects in any of many different cues can disrupt it, resulting in a failure to enter puberty, and infertility. This is associated with anosmia, a defect in the sense of smell, and the combination of symptoms is known as Kallmann syndrome, after Franz Josef Kallmann, who recognized its hereditary nature in 1944.[4]

Before GnRH was discovered, the first significant attempt to locate the source of the releasing factor regulating LH secretion was made by Halász and Pupp in 1965.[5] When the pituitary is separated from the hypothalamus, LH secretion is abolished; in males the testes shrink, and females stop ovulating. Using a cleverly designed microscalpel, Halász and Pupp cut all around the mediobasal hypothalamus of rats, isolating it from the rest of the brain. Surprisingly, the drastic surgery had no effect at all on testis weight or structure, but it blocked ovulation in females. This suggested that the LH pulses needed to maintain the testes were regulated by the mediobasal hypothalamus, but the LH surge needed for ovulation was probably regulated elsewhere. But we now know, as Halász and Pupp did not, that the mediobasal hypothalamus of rats contains the nerve terminals of GnRH neurons but few if any of the cell bodies. However shaky the foundations of their hypothesis were, it is a hypothesis that now looks correct. But before we come to that, we need to understand more about the observations that guided our present understanding, and particularly the studies of Ernst Knobil at the University of Texas.

To generate a pulse of GnRH secretion, many of the GnRH neurons must be activated at about the same time, probably firing a burst of spikes. In conscious rhesus monkeys, Knobil and his coworkers began recording multiunit activity—signals from the combined spike activity of many neurons. In the mediobasal hypothalamus, they observed occasional "volleys" of multiunit activity, bursts of spikes that lasted for several minutes. These volleys were invariably associated with LH pulses.[6,7]

This finding was astonishing. There are a few GnRH neurons in the mediobasal hypothalamus of the monkey, but other neurons in the same region regulate pulsatile secretion of growth hormone, prolactin, and thyroid-stimulating hormone, and no volleys were associated with any of *those* pulses. Equally unexpectedly, just before the LH surge there was a *reduction* in the frequency of volleys, followed sometimes by a complete cessation for a day or two—nothing was seen concurrent with the surge itself.

Similar observations were soon made in goats and in rats.[8,9] Although anesthetics generally block LH pulses, Kevin O'Byrne and his colleagues in King's College London showed that the pulses are intact in ketamine-anesthetized rats. As in monkeys, electrodes in the mediobasal hypothalamus could record volleys of multiunit activity that coincided with LH pulses. The volleys were very like those in rhesus monkeys: they lasted between one and four minutes, and occurred every twenty minutes, in accord with the higher frequency of LH pulses in rats.[10]

Knobil had thought it likely that the volleys came from GnRH neurons themselves; if not from the cell bodies, then from bundles of their axons. The volleys had a shape reminiscent of the bursts in vasopressin cells, and the mechanisms underlying those bursts were well understood. Did similar mechanisms underlie what Knobil called the "LH pulse generator"?

Many neurons generate bursts of spikes, including dopamine and NPY cells of the arcuate nucleus and mitral cells of the olfactory bulb; and so do many endocrine cells, including corticotrophs and gonadotrophs of the anterior pituitary and insulin-producing cells of the pancreas. In these different cells bursts are generated by different mechanisms, and the bursts have distinctive temporal features. The first bursting neuron to be studied extensively was neuron R15, a neuron in the abdominal ganglion of the mollusc *Aplysia californica* that is important for egg-laying.[11] R15 displays *parabolic bursting*, its bursts wax and wane, and bursts can develop without any synaptic input. Its membrane potential oscillates between a depolarized "up state" and a hyperpolarized "down state" because of a calcium channel that is activated when R15 is depolarized and inactivated when the intracellular calcium concentration is high. Bursts of spikes "ride" on the up state, and synaptic input modulates their intensity.[12]

Do GnRH neurons generate bursts? The ability to record their electrical activity was advanced by transgenic mouse lines in which the GnRH neurons had been engineered to express a fluorescent "reporter." In brain slices, kept alive in oxygenated medium, the GnRH neurons fluoresce, and microelectrodes can be introduced into them to study their properties. A few

GnRH neurons showed regular parabolic bursting: the bursts were shorter and more frequent than in R15, but, as in R15, they waxed and waned. However, while most neurons showed some irregular bursting, parabolic bursting was rare. None of the observed bursting conspicuously helped to understand LH pulses: the bursts were too frequent and mostly contained just a few spikes.

Other evidence came from immortalized GnRH cells (*GT1 cells*).[13] These were established by introducing, into mice, a transgene containing the promotor region of the GnRH gene coupled to the coding region of a tumor-promoting oncogene. This produced a tumor of the anterior hypothalamus, and the cells were dissociated and cloned. Some of these cell lines produced lots of GnRH, and GT1 cells secreted GnRH in regular pulses. Many of the genes that these tumor cells express are not the same as in "normal" GnRH neurons, but it was intriguing that these cells could generate pulses. In one of the first electrophysiological studies of GT1 cells, Charles and Hales saw that many of them showed regular rises in intracellular calcium at intervals that varied between 3 and 120 seconds in different cells, but the rises happened at different times in different cells.[14] Some cells showed occasional bursts of four or five spikes, but most spiked apparently randomly. However, longer recordings indicated that, although the bursts were not synchronous, the mean activity level of the cells fluctuated with a period of about 20 minutes, close to the period of GnRH pulses measured from similar cultures. These episodes seemed to arise from weak, spike-dependent signals.[15] Thus GT1 cells had *some* ability to generate regular rhythmic behavior, although the calcium oscillations could not explain GnRH pulses.

In 1999, Ei Terasawa and colleagues at the University of Wisconsin cultured neurons from the olfactory placode of rhesus monkey embryos, and saw that they secreted GnRH in pulses every hour or so—similar to the period of LH pulses in monkeys.[16,17] Many of the cultured GnRH neurons also showed oscillations in intracellular calcium but with a much shorter period, of about 8 minutes. This was longer than that of the calcium oscillations in GT1 cells but still too short to explain GnRH pulses. However, the oscillations were *intermittently* synchronized: every hour or so, many of the GnRH neurons showed synchronous rises of calcium. Similar reports followed from GnRH neurons cultured from embryos of mice, rats, and sheep.[18]

The calcium oscillations in GT1 cells had seemed to be only weakly linked with spike activity, and the same seemed to be true of the GnRH neurons from monkey embryos. Abe and Terasawa found that most fired very irregularly, and the only bursts were rather unimpressive, comprising clusters of at most ten spikes.[19] These are in marked contrast to the intense milk-ejection bursts of oxytocin cells, which typically contain more than a hundred spikes. Moreover, the timing of the bursts in GnRH neurons appeared to be random. In just 2 of 20 neurons was there a regular rhythmic pattern of bursts, but even these had at most seven spikes in each burst.

In many cells, oscillations in calcium arise not from spike-triggered calcium entry but from intracellular signaling pathways.[20] In embryonic mouse GnRH neurons at early stages of culture, calcium oscillations depend on spike activity: they are blocked by tetrodotoxin, which blocks the sodium channels on which spikes depend.[21] In older cultures, tetrodotoxin does not completely block the oscillations; these now involve mobilization of intracellular calcium. Perhaps only some calcium oscillations reflect spiking activity, and only these trigger the release of signals that affect other GnRH cells. It's not easy to tell: it is possible to measure intracellular calcium simultaneously in many neurons in culture, but not possible to measure spiking activity in many neurons simultaneously.

Nevertheless, these studies suggested that GnRH neurons interact in a way that leads to intermittent synchronization. GnRH cell bodies are not clustered together, but this does not mean that they do not contact each other. Rachel Campbell in Otago filled mouse GnRH neurons with a dye, biocytin, and discovered that the neurons are *not* anatomically isolated: they have long dendrites that come into close contact with each other,[22] and where the dendrites of two neurons intertwined, afferent axons often made synapses with both neurons.[23] Campbell speculated that these "shared synapses" might be important for synchronizing the electrical activity.

This seemed to put the responsibility for coordinating GnRH neurons onto neurons that were the origin of the shared synapses. Did the close contacts between GnRH neurons allow some signals to pass between them? The contacts are not synapses, nor do they involve gap junctions that might mediate electrical coupling. Could there be any chemically mediated interactions? Magnocellular oxytocin neurons also have dendrites that are bundled together. In nonlactating rats, these dendrites are sheathed in glial processes that insulate each from its neighbors, and the processes retract during lactation, bringing the dendrites close together. Accompanying this retraction, shared synapses appear, as in GnRH neurons.[24] However, these are mostly inhibitory, and while their presence means that there is some common input to adjacent neurons, that might still be only a very small part of the input. What is more important for oxytocin cells is that the dendritic contacts facilitate chemical interactions: the dendrites contain many vesicles, some of which are released during suckling, and the oxytocin that is released acts on oxytocin receptors expressed by oxytocin cells.

Both GT1 cells and embryonic GnRH neurons express GnRH receptors, and in these and in adult GnRH neurons GnRH is excitatory,[25] but the dendrites have few large vesicles. Thus, although it seems that GnRH neurons must intercommunicate, how much of this is direct and how much is mediated by other neurons is unclear. Linked to this is the issue of how the bursts of activity that underlie pulsatile GnRH secretion arise—whether GnRH neurons generate these bursts themselves, or respond to a bursting input from another cell population.

Neuroscientists prefer to study neurons in vitro or in cell cultures rather than in living animals, in part from ethical concerns, in part to take advantage of simpler preparations and the opportunities for more precise control of experimental conditions. However, there are problems with interpreting electrophysiological recordings made in vitro. In brain slice preparations, axons and dendrites are severed, so most synaptic inputs are lost, and this can radically alter the behavior of neurons. Unfortunately, the scattered distribution of GnRH cells has made it almost impossible to study them in living animals. The only recordings so far, made by Stephanie Constantin,[26] were in conditions incompatible with simultaneously measuring LH. Those recordings showed that GnRH neurons are very heterogeneous, but they revealed little in the way of behavior that could be related either to the multiunit activity recorded by Knobil and others, or to the activity of GT1 neurons, or to that of GnRH neurons in vitro.

The best that electrophysiological studies in vitro can do is display the repertoire of mechanisms available to GnRH neurons that might contribute to their behavior in the living animal. However, the repertoire is vast, and different elements of it are available to different GnRH neurons. This is not unique to GnRH neurons. Even though we might be studying cells of a tightly defined population isolated from many factors that, in their normal environments, cause differences in their behavior, we still see great heterogeneity. Sometimes, we see *more* heterogeneity when we isolate neurons, because cells adapt to their environment. Neurons that receive more or less of a particular chemical signal typically respond by down-regulating or up-regulating expression of receptors for that chemical. Neurons that receive a stronger or weaker excitatory drive often alter their expression of ion channels to restore a "normal" level of excitability. When we isolate neurons, we see differences that in part reflect adaptations to their previous individual histories.

If bursts are important, there might be many different mechanistic pathways to that bursting. They don't necessarily arise from intrinsic properties: growth hormone is secreted from the anterior pituitary in large pulses every few hours. These pulses arise because the somatotrophs of the anterior pituitary are alternately stimulated by GHRH from the arcuate nucleus and inhibited by somatostatin from the periventricular nucleus: the somatostatin neurons are activated by delayed effects of growth hormone secreted into the blood, and they inhibit the GHRH neurons as well as inhibiting growth hormone release. There are many means by which bursts can be generated and many more by which pulses arise, and often, it seems, bursts and pulses are generated by multiple, overlying mechanisms, involving both intrinsic properties and network mechanisms.

In any complex biological system subject to natural selection, many mutations will have accumulated over the long and erratic course of evolution, a course that takes no straight path from sufficiency to excellence, but which weaves through an ever-changing landscape. Mutations that impair a behavior that at some place in that landscape is critical may be eliminated, but any that duplicate or mimic some existing facet are likely to persist. This type of redundancy—better known as *degeneracy*—arises spontaneously and inevitably, and it supports the robustness of biological systems. To the researcher, it presents difficulties. It's hard to design experiments to test the importance of a mechanism when neurons can adapt to its failure.

Degeneracy in biological systems can also lead naturally to differences between species. Perhaps the only "law" in biology is that *there are (almost) always exceptions*. In horses, it is possible to sample the blood that enters the pituitary gland from the hypothalamus by introducing a cannula into the facial vein and pushing it into the cavernous sinus that lies above the pituitary gland. Samples taken this way show, as expected, that pulses of LH secretion are invariably preceded by GnRH pulses. However, in mares there is *no* preovulatory LH surge, only a modest increase in the frequency of LH pulses—but this is enough to trigger ovulation, given the dramatic increase in LH receptor expression in the dominant follicle. The increase in LH pulse frequency is not a stimulatory effect of estrogen on LH secretion; in the mare, estrogen levels begin to fall *before* ovulation, not as a result of it, and it is this that leads to the progressive acceleration of GnRH and LH secretion.[27]

What is currently known about the properties of the GnRH neurons in the mammalian hypothalamus seems of little help in understanding the generation of pulses. Their observed electrical activity has at best a tenuous relationship with the pulse-generator volleys. GnRH neurons are heterogeneous, and are not tightly coupled together. Some signals probably pass among them; those signals might include GnRH, but GnRH signaling seems not to be essential. However, looking for the explanation of pulses in the properties of GnRH neurons was always a long shot. These neurons are embedded in a network of many different cell types. Nearby GABA neurons appeared to be important, as did distant noradrenergic inputs from the caudal brainstem, but GABA and noradrenaline seem to be important for everything and nothing, depending on how you look. Further progress in understanding how the hypothalamus regulates LH secretion required some new discovery: that came, as described in the next chapter, with the discovery of a new peptide—kisspeptin.

9 Kisspeptin

Was this the face that launch'd a thousand ships,
And burnt the topless towers of Ilium?
Sweet Helen, make me immortal with a kiss.
—Christopher Marlowe (1564–1593), "The Face That Launch'd a Thousand Ships"

By the end of the 1990s, reproductive neuroendocrinology was confronted with seemingly intractable puzzles. The most obvious was that, while the ovarian hormones estrogen and progesterone were immensely important for GnRH secretion, we were unsure about how or where they acted.

While estrogen depresses pulsatile LH secretion, high levels of estrogen are an essential predeterminant of the preovulatory LH surge.[1] Pulsatile secretion seems to be regulated by the arcuate nucleus, where the axons of GnRH neurons enter the median eminence, but surges seem to be controlled in the rostral hypothalamus where most of the cell bodies of the GnRH neurons are—at least, most of those that control the pituitary.

There are three known estrogen receptors: *estrogen receptor alpha* (ERα), produced by the gene ESR1; *ERβ*, produced by the gene ESR2, and a *G protein-coupled estrogen receptor*. ERα and ERβ are *nuclear receptors*: their "classic" actions are not directly on neuronal excitability as those of membrane receptors are, but on gene expression. Estrogen is lipid soluble and so it can freely enter cells. If either ERα or ERβ is expressed by a cell then estrogen will bind to it and the bound complex is translocated to the cell nucleus where it interacts with *estrogen response elements* in the DNA, altering gene expression. Some GnRH cells express ERβ,[2] but this receptor is not needed for the effects of estrogen on LH secretion.[3] Estrogen inhibits GnRH pulses via ERα, but GnRH neurons do not express it.[4]

We thus needed to find a population of ERα-expressing neurons to explain the GnRH surge; perhaps they were in the rostral hypothalamus, where many estrogen-receptive cells are found close to the GnRH neurons. If those cells initiate a surge of spiking in the GnRH neurons that leads to the LH surge, then the multiunit recordings of Knobil and others probably did not come from the axons of GnRH neurons, because no multiunit activity is seen in conjunction with the surge.

We also needed to find a population of ERα-expressing neurons to explain how estrogen inhibits GnRH pulses. If Knobil's LH pulse generator reflects neurons in the arcuate nucleus that regulate GnRH neurons, then perhaps those express ERα. Is it plausible that neurons in the arcuate nucleus generate bursts of spikes that are propagated along axons that end on the cell bodies or dendrites of the GnRH neurons far away in the rostral hypothalamus, which then send axons back through the arcuate nucleus to end in the median eminence? This seemed perverse. It seemed more likely that pulse-generator neurons in the arcuate nucleus would interact directly with the GnRH axons in the arcuate nucleus. But all that synaptic inputs to axons were known to do was to inhibit spike activity—classically, spikes are initiated close to the cell body.

A breakthrough came with the discovery of kisspeptin. Kisspeptin is a peptide encoded by the *Kiss1* gene, identified by scientists in Hershey, Pennsylvania, as a gene that could suppress melanoma and breast cancer metastasis.[5] They named it after Hershey's Kisses, the popular American chocolate-teardrop candy—sadly, not for any link with reproductive behavior.

In 2003 it was reported that mutations of the kisspeptin receptor were associated with hypogonadotropic hypogonadism in humans—in this condition, impaired secretion of LH and FSH leads to dysfunction of the testes and ovaries, and puberty is delayed or absent.[6,7] Kisspeptin does not affect the secretion of LH and FSH directly, but stimulates GnRH release. It is expressed by two main populations of neurons, one in the arcuate nucleus and a more rostral population. Both express the critical estrogen receptor ERα.

The kisspeptin neurons of the rostral hypothalamus are particularly prominent in females. This is because, in these neurons, *Kiss1* expression is upregulated by estrogen. These neurons are essential for the preovulatory LH surge, which can be blocked by infusion of an antibody to kisspeptin into the rostral hypothalamus. By contrast, in the arcuate nucleus, *Kiss1* expression is *down*regulated by estrogen, and these neurons are critical for LH pulses but not for the LH surge. After the menopause, when the ovaries produce no more steroid hormones, pulsatile secretion of LH in women is exaggerated.

Most of the kisspeptin neurons in the arcuate nucleus express two other peptides, neurokinin B and dynorphin, and hence they are sometimes called KNDy neurons (kisspeptin neurons in the rostral hypothalamus contain neither neurokinin B nor dynorphin). In humans, mutations of either *TAC3*, which encodes neurokinin B, or *TAC3R*, which encodes its receptor, are associated with gonadotropin deficiency and failure to reach puberty.[8] Transgenic mice that lack either dynorphin or the kappa opioid receptors through which dynorphin acts also have deficiencies in LH secretion. Thus neurokinin B and dynorphin are also important for GnRH secretion.

The KNDy neurons talk to each other. They express receptors for neurokinin B and dynorphin, they are surrounded by axons that contain all three peptides, and axons containing neurokinin B connect one arcuate nucleus with the arcuate nucleus of the other side of the brain. Whether kisspeptin itself regulates KNDy neurons is less clear, but it certainly acts on GnRH neurons. In the monkey, kisspeptin axons are densely intermingled with GnRH axons in the median eminence,[9] and pulses of kisspeptin release occur in conjunction with pulses of GnRH.[10]

Transgenic mice that lack kisspeptin receptors are infertile, and restoring kisspeptin receptors in just the GnRH neurons completely rescues their fertility. In Otago, Allan Herbison set about testing whether KNDy neurons could produce pulsatile LH secretion. For this he used *optogenetics*, which involves making neurons sensitive to light by introducing a gene that encodes a light-sensitive receptor, such as *channelrhodopsin2*.

Channelrhodopsin2 is normally expressed in green algae; it is an ion channel that guides the movement of the cells in response to light—the algae rise to the surface of a pond by day and sink at night. In 2005, Karl Deisseroth's lab showed that modified lentiviruses could be used to introduce channelrhodopsin2 into neurons, whose spike activity could then be controlled by exposing them to light of the appropriate wavelength.[11] Lentiviruses are retroviruses, which integrate some of their RNA into the DNA of cells that they infect. Now, many transgenic mouse lines are available where many kinds of light-sensitive channels have been introduced into specific neuronal populations, and virus-based approaches have made it possible to easily introduce these channels into specific cell populations in rats.

Allan Herbison and his colleagues introduced channelrhodopsin into the KNDy neurons and placed an optic fiber into the hypothalamus, enabling them to activate the KNDy neurons with flashes of light.[12] When they activated the KNDy neurons in male mice or in ovariectomized female mice, making them fire at 10 spikes per second for two minutes was enough to generate an LH pulse. In diestrous female mice, a higher frequency (20 spikes per second) was needed; because there is less kisspeptin in the KNDy neurons at this stage of the ovarian cycle, the neurons must be stimulated harder to achieve the same effect.

However, puzzles remained. Microinjecting kisspeptin either into the rostral hypothalamus or into the arcuate nucleus could stimulate LH secretion, but whereas injecting a kisspeptin antagonist into the arcuate nucleus suppressed LH pulses, rostral injections had no effect on them.[13] It thus seemed that, although kisspeptin was an essential driver of the GnRH release that drives LH pulses, it was not acting at the cell bodies of GnRH neurons, but close to the axon terminals. How could an action at that site, far away from the presumed origin of spikes in GnRH neurons, trigger pulses that were presumed to be the result of burst of spikes in GnRH cells?

What had been thought to be the axons of GnRH neurons do not seem to be axons as conventionally understood, but long dendrites from which a cluster of short axons emerges at the level of the median eminence. Herbison called these *dendrons* to suggest a hybrid of dendritic and axonal properties: the dendron does not merely transmit spikes generated at or close to the cell body—it receives signals along its whole length. Iremonger and Herbison proposed that the GnRH neuron is a hybrid of endocrine cell and neuron and that, at the terminals, kisspeptin controls GnRH secretion not by triggering spikes, but by modulating calcium entry into the distal end of the dendron.[14]

However, to get enough calcium entry to trigger secretion generally requires calcium channels that are opened only at the voltages reached by spike activity. So might spikes be generated in a dendron? Spikes can be initiated anywhere along an axon, given enough depolarization: all that makes a spike-generating site unique is a high density of voltage-gated sodium channels. In many neurons, dendrites have voltage-gated sodium channels and can propagate spikes; in these neurons, spikes are normally initiated close to the cell body and are propagated along the dendrites as well as along the axons. However, in some neurons, spikes can also be initiated in the dendrites; so perhaps inputs from kisspeptin neurons might trigger bursts of spikes at the end of the dendron. Perhaps the GnRH neurons have *two* sites of spike initiation: one at the end of the dendron where inputs from KNDy neurons generate the bursts that drive pulsatile secretion, and another, close to the cell bodies, where neurons of the rostral hypothalamus generate a surge of spike activity that produces the preovulatory LH surge[15] (figure 9.1).

The discovery of kisspeptin has not answered all our questions, but has posed many of them in a new way. It seems likely that bursts are generated by arcuate kisspeptin neurons that signal to the dendrons of GnRH neurons. To understand the generation of those bursts we must begin afresh by studying the arcuate KNDy neurons, which seem to have intrinsic properties compatible with burst firing.[17] Bursts can be generated in many different ways, and the kind of properties that we expect of burst-generating neurons are the subject of the next chapter. The answer to the GnRH surge may lie in the rostral kisspeptin neurons, but it has yet to be found: as of now we know little of their properties. The significance of the calcium oscillations in GnRH neurons is still a mystery.

This is what science is like—every discovery raises new questions, often forcing us to reconsider things we have taken for granted. This summary has scarcely begun to encompass the disputes, conundrums, and paradoxes of GnRH research. We respond to new discoveries by trying to incorporate them into our existing body of knowledge. We try to explain as much as possible: we don't abandon old evidence, though we might look at it a new way. Hypothesis testing as envisaged by Karl Popper is part of this process. Generally, hypotheses are not grand and universal bold hypotheses: they are answers we offer for puzzles posed by observations that we can't easily dismiss.

I began by emphasizing how few GnRH neurons there are. Without these few GnRH neurons, we cannot reproduce, and our own lines must end; it is a cause for pause to consider how much rests on so few. Yet, there seem to be many more than are strictly needed. Normal mice have about 600 GnRH neurons in the hypothalamus. Their migration is impaired in the mutant mouse strain GNR23, and in homozygous mice only about 70 reach the hypothalamus. Males of this strain have smaller testes than

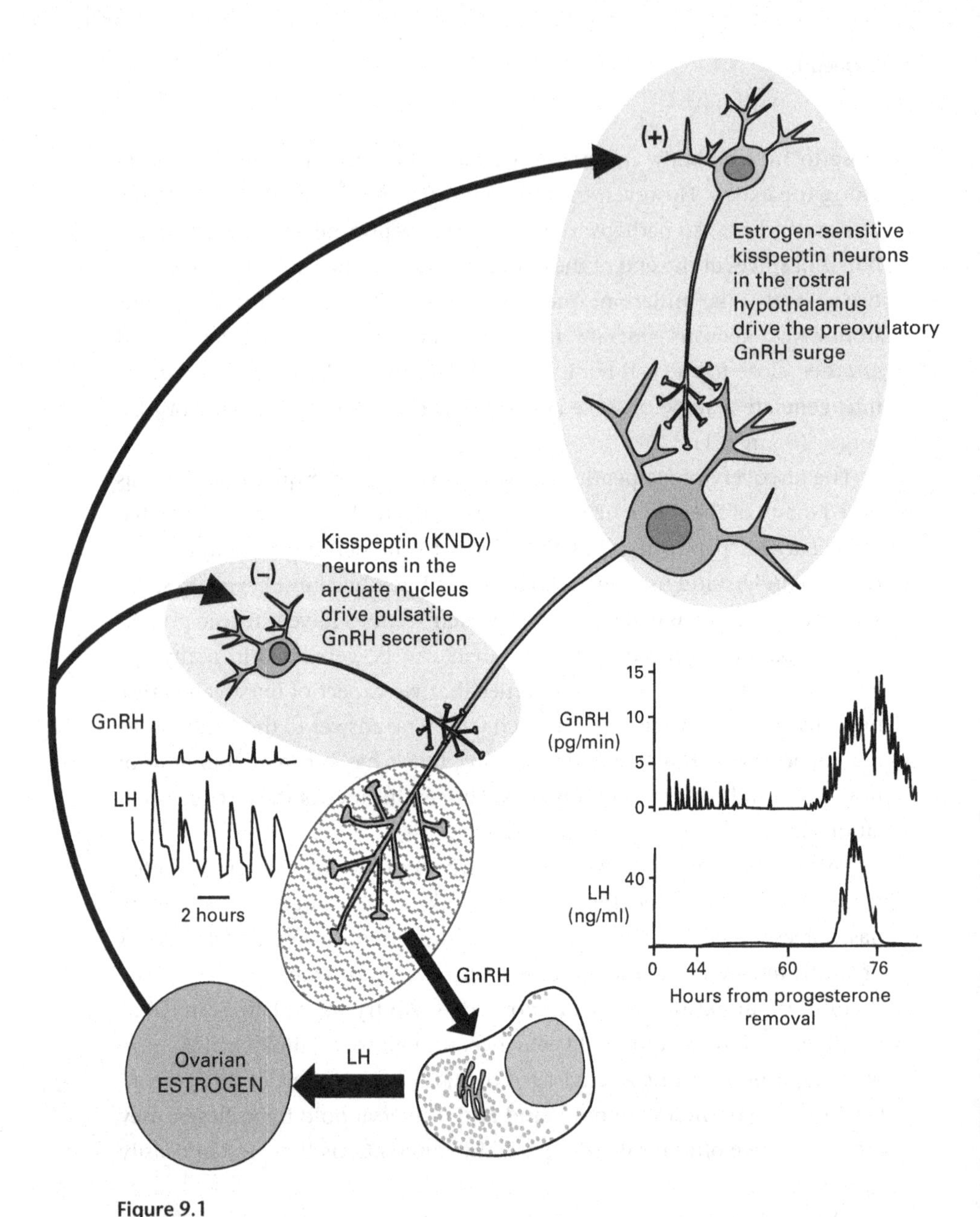

Figure 9.1

Kisspeptin and the GnRH neuron. The insets illustrating pulsatile secretion of GnRH and LH on the left and surge secretion on the right are adapted from published work of Sue Moenter, Alain Caraty, Alain Locatelli, and Fred Karsch, showing that "GnRH secretion leading up to ovulation in the ewe is dynamic, beginning with slow pulses during the luteal phase, progressing to higher frequency pulses during the follicular phase and invariably culminating in a robust surge of GnRH."[16]

normal, but are fully fertile; females enter puberty normally but are generally infertile. Mice with just one copy of the mutant gene have about 200 GnRH neurons in the hypothalamus, and these females are fully fertile with normal ovarian cycles.

Thus very few GnRH neurons seem to be needed.[18] But we cannot be sure. Aging is accompanied by a decline in the production of GnRH. In 2013, a paper in *Nature* suggested that the decline of GnRH contributes directly to the process of aging and that life span itself can be extended by maintaining the population of GnRH neurons.[19] It would be unwise to put too much store on one paper; the scientific literature is full of false leads and dashed hopes, but it is an enticing thought that one day we might indeed be made immortal, and perhaps even with a kiss.

10 The Bistable Neuron

The Grand Old Duke of York
He had ten thousand men
He marched them up to the top of the hill
And he marched them down again
And when they were up they were up
And when they were down they were down
And when they were only half way up
They were neither up nor down.
(Traditional)

The traditional nursery rhyme mocks the vacillation and indecision that supposedly led the Duke of York's forces to defeat at the hands of a lesser force of Napoleonic revolutionaries at Tourcoing in Flanders in 1794. In the rat, there are about 8,000 magnocellular vasopressin neurons, and each seems to be as futilely indecisive as the Duke of York.

When Wakerley and Lincoln recognized oxytocin cells by their bursting discharge in response to suckling, they also, as a by-product of the discovery, recognized vasopressin cells. Because all neurons in the supraoptic nucleus make either oxytocin or vasopressin, the neurons *not* activated by suckling had to be vasopressin cells. Many of these fired phasically, with long bursts separated by long periods of silence (figure 10.1).

If we knew nothing else about these cells, we might think that each codes a massive amount of information in this pattern. We might imagine that each spike is a "word" whose meaning depends on its timing relative to the previous spike, and that each burst is thus a "sentence" in a stream of communicated information. This analogy underlies common calculations about the information coding capacity of the brain; for vasopressin cells it could hardly be more wrong.

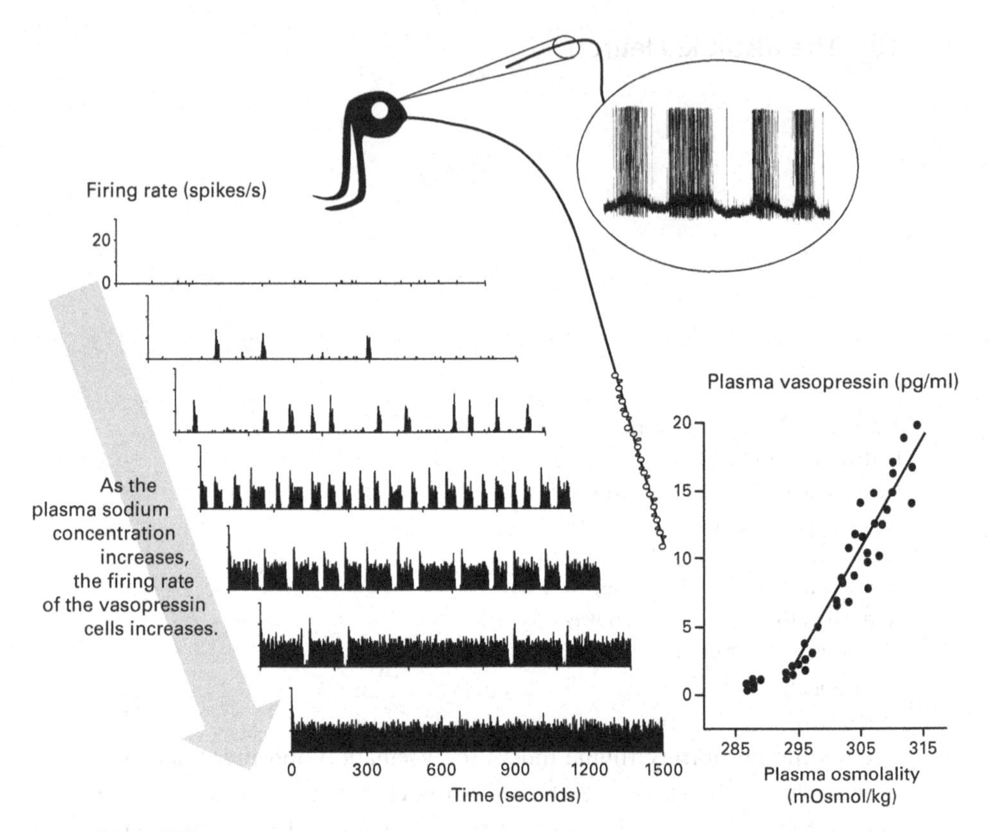

Figure 10.1

Phasic firing in vasopressin neurons. Magnocellular vasopressin neurons secrete vasopressin into the blood from the swellings and nerve endings of axons in the posterior pituitary gland. This secretion is governed by the spike activity of the neurons. The spike activity can be monitored by a microelectrode whose tip is placed either inside a vasopressin cell (intracellular recording) or just outside the cell (extracellular recording). These neurons discharge spikes in long bursts of activity separated by long silences. Intracellular recordings made in vitro have revealed the mechanisms underlying these bursts, and one early key observation, shown here, is that the bursts of spikes "ride" upon a plateau of depolarization that is the result of a depolarizing afterpotential.[1] The mechanisms have been reconstructed in computational models of vasopressin cells, and the traces on the left are simulations of the activity of a vasopressin neuron in response to increasing osmotic stimulation.[2] The graph on the right shows measurements of vasopressin in rat blood, and it shows how the vasopressin concentration is linearly related to the plasma osmotic pressure.[3]

For any particular vasopressin cell, the pattern of bursts is quite consistent, but while one cell might fire in bursts that last 20 seconds or so at about five spikes per second, its neighbor might fire at ten spikes per second in bursts that last 10 or 60 seconds. The cells fire asynchronously, so the bursts do not generate a pulsatile pattern of secretion from the pituitary. Unlike most pituitary hormones, vasopressin is secreted continuously into the blood from the discordant cacophony of neurons firing erratically and independently. The *pattern* of spiking in any individual cell has apparently no meaning whatsoever.

Nevertheless, given a challenge that is not slow and continuing but sharp and sudden, perhaps the bursts might be synchronized to generate a pulse of secretion. Not so.

Early studies investigated the effects of injecting agents like phenylephrine that caused a large, brief rise in blood pressure. These (generally) inhibited vasopressin cells, seemingly consistent with a role in blood pressure regulation. This seemed to make sense—vasopressin raises blood pressure by constricting blood vessels, so vasopressin secretion when blood pressure falls would seem to be a useful homeostatic response, but, in the rat, vasopressin cells are not very sensitive to arterial blood pressure: modest sustained changes have little effect. Much more important are the low pressure venous receptors that signal plasma volume. It seems that vasopressin is important not for maintaining a steady arterial blood pressure, but for maintaining a constant pressure in the veins.

Looking at the responses to phenylephrine, Richard Dyball and I noticed that phasic neurons were not *always* inhibited by phenylephrine; it depended on when the injection was given. Toward the expected end of a burst, an injection would stop a burst, but if it was given just after a burst had started then often the burst would be prolonged.[4] This made no physiological sense at the time, and we left it aside for another day. But it did make some sense to my mathematical intuitions.

My time as a student of mathematics was interspersed with rare moments of enlightenment. One came when I understood what it meant for an equation to have two or more solutions. Of course I knew, as a matter of fact, that an equation could have two solutions, that $x^2=1$ could be solved by $x=1$ and by $x-1$, but there is a difference between knowing something as a matter of fact and knowing its meaning. For me, the meaning became apparent when I learned about the Belousov-Zhabotinsky reaction.[5] In the first described example, potassium bromate, cerium sulfate, malonic acid, and citric acid are mixed with dilute sulfuric acid. As the mixture is stirred, ceric ions appear and disappear periodically; accompanying changes in light absorption cause the solution to oscillate between yellow and colorless, and a platinum electrode dipped into the solution will record oscillations in voltage. Thus a simple combination of chemicals could show repeated, rhythmic alternations between two states.

The reactions can be represented as differential equations, and similar equations can be used to describe the bursting behavior of neurons. A differential equation describes how a variable, v, perhaps the spiking activity of a neuron, changes with time as a function of other variables. The rate of change of v is denoted by dv/dt, and the equation

$$dv/dt = -v\,(v - C)\,(v - P) \text{ where } 0 < C < P$$

defines a *dynamical system* that has two stable states, corresponding to $v = 0$ and $v = P$. These states are *stable* because if v moves slightly away from 0 or from P as the result of a small perturbation, then the equation dictates that the value of v will return to either 0 or P. If v starts at 0 and is strongly perturbed, to $v > C$, then v will not return to 0 but will increase further to P. Conversely if v starts at P and is inhibited to $v < C$, then v will decline to 0. C defines an *unstable equilibrium* between the two *stable equilibria* of 0 and P.

It is possible to make a simple model of a bursting cell from these beginnings.[6] To do so, we need to add a variable w that defines a slow activity-dependent inhibition, and can write the equations as

$$dv/dt = s\,[-v\,(v - C)\,(v - P) - k_1 w]$$

and

$$dw/dt = b\,(k_2 v - k_3 w)$$

The term $k_2 v$ defines w as increasing in response to activity, while $k_3 w$ determines that, in the absence of activity, w will decline to zero (b, s, k_1, k_2, and k_3 are just scaling factors). These equations define a dynamical system with two stable states corresponding to a burst of spike activity and inactivity, and define a process whereby the burst will eventually end. When v is subjected to continual small random perturbations, the system can alternate between long bursts and long silences, superficially like vasopressin cells (figure 10.2).

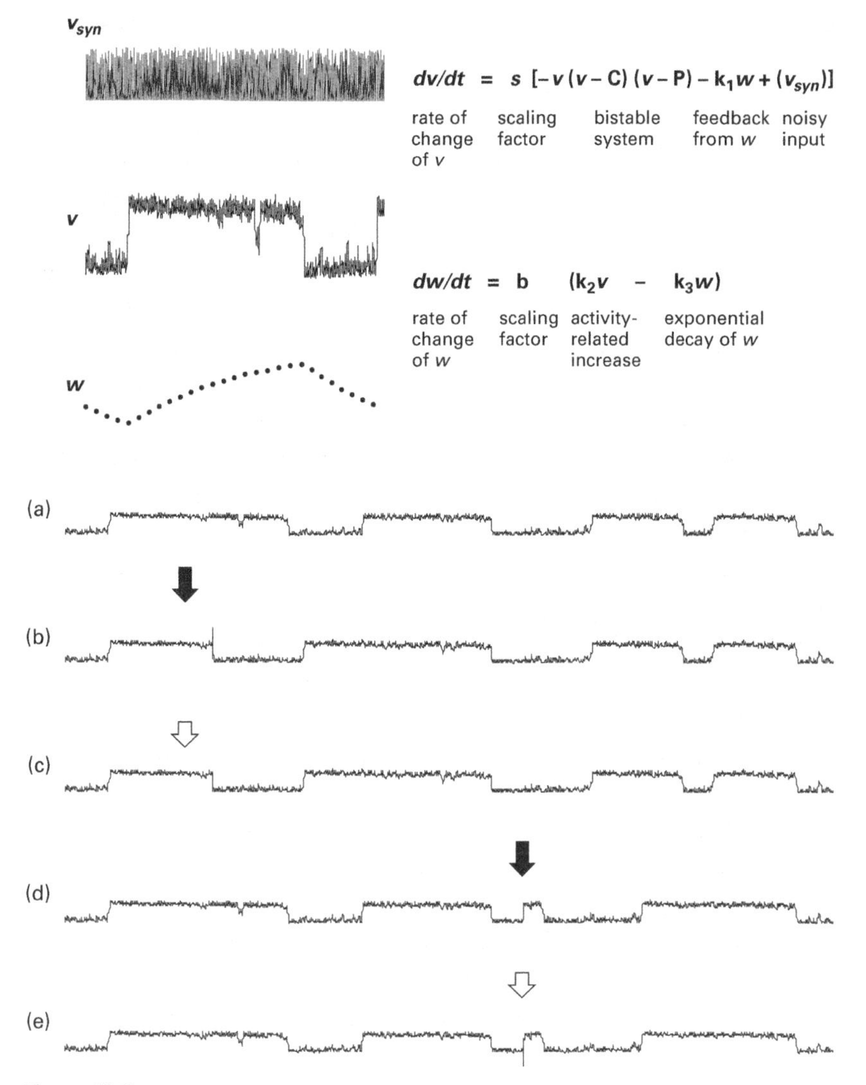

Figure 10.2

Bistability. *Bistability* in neurons can arise from a positive feedback mechanism that saturates: spike activity in vasopressin cells produces a depolarizing afterpotential that increases neuronal excitability, but its effect saturates because short hyperpolarizing afterpotentials after each spike limit how fast the neuron can fire. *Bistable oscillations* arise because of a slower activity-dependent inhibition—the *afterhyperpolarization*. This can be mimicked by two differential equations challenged by a "noisy" input mimicking synaptic input. Here, v represents neuronal activity and w represents activity-dependent inhibition; this system alternates between "bursts" and "silences." In such a system, perturbations can have paradoxical effects. (a) shows the system "firing phasically" in response to a constant, noisy input (v_{syn}). (b) Here, the input is the same as in (a), but at the arrow a brief additional excitation *stops* the first burst. (c) The same effect occurs with a brief inhibition (open arrow). (d) Here, a brief excitation given in a silent period triggers a burst, and (e) a brief inhibition at the same time also triggers a burst.

To make a more realistic model, we need to understand how a cell can have two stable states. A vasopressin cell isolated from synaptic inputs will remain at its resting potential, and this is one obvious stable state. In a living animal, a vasopressin cell continuously receives inputs that cause many small perturbations of the membrane potential. Thus the membrane potential is continually fluctuating, and when a fluctuation exceeds the spike threshold, a spike will be triggered. That spike will trigger entry of calcium, producing a sequence of changes in neuronal excitability. After a spike, a *hyperpolarizing afterpotential* means that the cell cannot fire again for at least about 20 milliseconds. As this spike decays, a *depolarizing afterpotential* makes the cell *more* likely to fire a spike, and if it does so then the depolarizing afterpotentials will summate, increasing the chance of yet another spike. Thus a burst begins. The burst is moderated by yet another spike-dependent potential, a *slow afterhyperpolarization*. These factors together will establish an approximate equilibrium—a "plateau potential," upon which spikes ride during a burst, and the burst will continue at a relatively steady rate of spiking.[6,7]

This burst will not easily stop once it has begun: to stop it, another activity-dependent process is required. As in oxytocin cells, the dendrites of vasopressin cells are filled with vesicles, and these contain not only vasopressin, but also dynorphin. Each vesicle contains about 85,000 molecules of vasopressin and only about 300 of dynorphin, but vasopressin cells have abundant dynorphin receptors, so dynorphin acts back on the cell that releases it. Vasopressin cells release vesicles from their dendrites reluctantly; they are powerful and precious cargo. But a long burst of spikes will cause some to be released, and when they are, as shown by Colin Brown and Charles Bourque, the dynorphin blocks the depolarizing afterpotential.[8] As a result, the burst stops, and for several seconds the cell will be inexcitable.

Duncan Macgregor showed that these properties, when expressed as differential equations and blended with equations that define how spikes are generated, produce a computational model whose output is indistinguishable from the spiking activity of vasopressin cells.[9] Vasopressin cells are not identical: they receive different inputs, and their membrane properties all differ slightly. But for any particular cell we can match its behavior closely by adjusting the parameter values of the model using a "genetic algorithm" that, by a process analogous to selection and recombination, efficiently finds the best fit.[10]

The properties of vasopressin cells define them as *bistable oscillators*: they can remain in either an active state or a quiescent state, but perturbations can "flip" them from one state into the other. An interesting prediction of the model is that when a vasopressin cell is firing a burst, a small and transient excitatory input can stop it, paradoxically having an inhibitory effect.[11] Whether real vasopressin cells behave in this way was tested by Nancy Sabatier. She recorded from phasic neurons while applying stimuli to the organum vasculosum of the lamina terminalis (OVLT), which supplies an excitatory input to vasopressin cells.[12] She set the intensity of stimulation low, to just perceptibly excite the vasopressin cells, and applied stimuli every five seconds. When a stimulus came shortly after a burst had ended, it had no effect, but when it came close to the time that a burst was expected, it would trigger a burst prematurely. When the stimulus came just after a burst had started, it had little effect, but when it came later in a burst, the cell would briefly increase its firing rate and then stop.

The results were clear; a brief excitatory input might either trigger a burst, or stop a burst, or have no effect. What was true in the model was true in real cells. This seems nonsensical; what sense can we make of a signal that can be either excitatory or inhibitory depending on the random state of its target? But we must think of what physiological functions the vasopressin cells fulfill: what matters for the kidney is the overall output from the whole population.[13] A stimulus that excites some neurons while inhibiting others will have little or no effect that matters.

The vasopressin system needs to respond to plasma sodium sensitively, stably, and accurately. These needs are in conflict. Any sensory receptor system, whether electrical, mechanical or biological, will be noisy at the limits of its sensitivity. In the sky on an unclouded night, the stars seem to twinkle. Stars don't twinkle: the twinkling is in the eye, in the response of a few noisy photoreceptors to the random arrival of photons from those so distant stars.

An increase in sodium concentration of the extracellular fluid that surrounds a cell produces in increase in osmotic pressure—a force that can draw water out from the cell. The vasopressin cells are themselves sensitive to these changes, as first shown by my colleague at Babraham, Bill Mason.[14] As osmotic pressure rises, the cells shrink slightly, and this affects stretch-sensitive channels in their membrane causing a depolarizing current to flow—a mechanism later revealed by Charles Bourque and his colleagues.[15]

However, this can only depolarize the cells by a few millivolts—too little to explain, on its own, the changes in spike activity that physiological increases in plasma sodium concentration produce. But in a living animal, vasopressin cells receive thousands of synaptic inputs, and some of these come from parts of the forebrain that are strongly implicated in thirst and are themselves osmosensitive.

To be maximally sensitive to changes in plasma sodium, a vasopressin cell must be sensitive to the rate at which signals arrive from osmoreceptors—and so it must be very sensitive to synaptic inputs per se, wherever they come from. Synaptic inputs come from many different sources and fluctuate randomly, so if a vasopressin cell is hypersensitive, then its activity too will be noisy. Hypersensitive neurons will not only be sensitive to neurons that carry osmotic signals, they will also be transiently perturbed by intrusive but irrelevant signals.

In fact, the extreme sensitivity of osmosensitive neurons *depends* on noise. Neurons are nonlinear devices: without synaptic input most remain quiet, at their resting potential. If given a gently increasing depolarization, they will remain quiet until the depolarization exceeds the spike threshold—several millivolts above the resting potential—at which point they will start to discharge spikes regularly. They will fire faster if the depolarization continues to increase, but their dynamic range is narrow—they will soon be firing as fast as they can. The maximum firing rate that vasopressin cells can sustain is limited by the hyperpolarizing afterpotential that follows each spike—in practice, they can't keep firing at more than about 10 spikes per second. This is not what we want of any sensory system: we want it to be as sensitive as possible and to have a wide dynamic range.

If the membrane potential of a neuron is constantly fluctuating around the resting potential, a small sustained depolarization, by altering the probability that fluctuations will exceed the spike threshold, will cause an increase in firing rate. This phenomenon, by which noise enhances the sensitivity of neurons, is called *stochastic resonance*.[16,17]

When we first proposed that this phenomenon underlies the osmotic responses of vasopressin cells, the idea was met with skepticism. If the OVLT and other forebrain structures are lesioned, then vasopressin secretion is low and will not rise when the plasma sodium concentration rises, and hence these sites had confidently been identified as the location of the osmoreceptors. Bill Mason and I challenged this: we recognized that the cells needed a noisy input to express that osmosensitivity, and so argued that perhaps the inputs from the forebrain merely provided a source of noise for the vasopressin cells.[18]

The idea that neurons might do something useful by providing noise was unpopular, and one skeptic was my friend and collaborator John Russell. Between us we devised a "killer experiment" to test it: we did the experiments together, and we had agreed that we would analyze the data together and publish the outcome whatever it was. In anesthetized rats, we lesioned the OVLT and other forebrain structures, and saw, as expected, that the vasopressin cells fell silent and were unresponsive to increases in plasma sodium. We then mimicked the lost synaptic input by trickling glutamate onto the cells to restore a low level of spiking activity. Then we again tested their responses to plasma sodium. The outcome was a judgment of Solomon: the cells *did* respond to plasma sodium, but not as strongly as in intact rats.[19] We were both right: the cells needed a tonic synaptic input, and with that they could respond to plasma sodium, and their response was enhanced by the forebrain structures, which therefore must also include osmosensitive neurons, as was soon confirmed by others.

So the vasopressin cells, by exploiting stochastic resonance, can be very sensitive to small changes in plasma sodium, and can, as we demonstrated with a computational model, increase their mean firing rate linearly over a wide dynamic range. Although the *mean* activity increases linearly, for any one cell the activity typically comprises a sequence of bursts. These bursts occur at different times in different cells, and the population of vasopressin cells as a whole averages out the noisy activity of all the cells to produce a steady signal for the kidneys (figure 10.3). If a transient perturbation arrives as a brief excitation to these cells, some will be excited, but because of the bistable properties of the cells, some will be inhibited: thus the system filters out irrelevant, transient perturbations. It acts as a low-pass filter: slow changes in plasma sodium are transmitted; erratic perturbations from second to second are not.

Estimates of the rate of information flow in neurons generally assume that spike timing is critical—that every spike is meaningful, and that the distance between spikes holds useful information. But this is not necessarily the case. In the days when I fixed my car myself I once replaced a rear

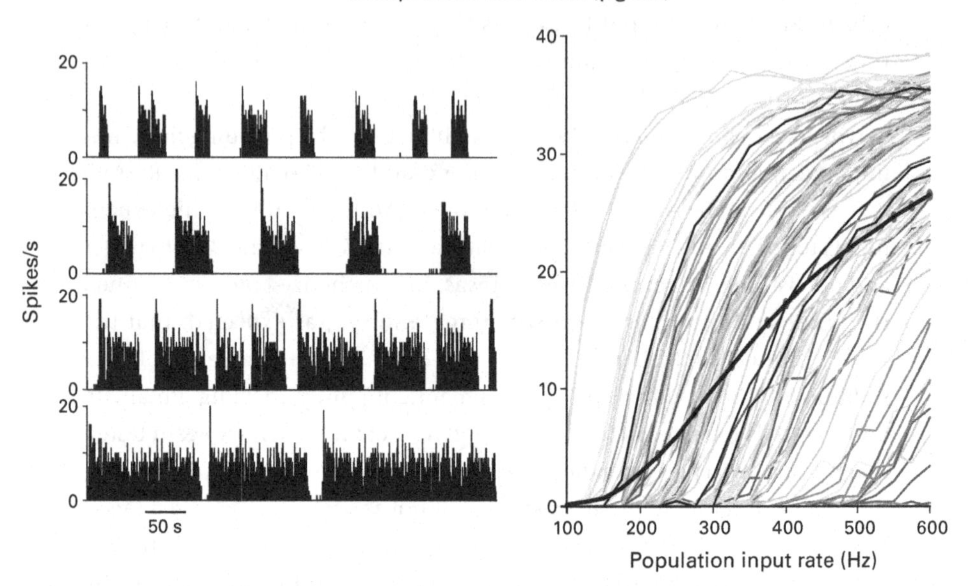

Figure 10.3
Heterogeneity. Vasopressin cells all fire in different patterns—the examples on the left show some of the different types of patterns in which different neurons fire at any given time. The neurons also vary in their mean firing rate and in their sensitivity to osmotic pressure. The mechanisms of phasic firing and the heterogeneity of neurons were modeled by Duncan MacGregor, who also modeled the pattern of vasopressin secretion from each cell. On the right is the modeled average secretion from 100 different vasopressin cells as a function of the average synaptic input to the population. Secretion from each one is very nonlinear, but the total secretion from the population, shown as the heavy black line, is linearly proportional to the input rate.[2]

indicator light, and asked a friend to check that it was working. "Yes," he said—"No, wait, no, yes, no, yes, no...." The flashing makes the signal more obvious, but the precise timing of the flashes is irrelevant.

The signals that neurons generate cannot be understood in isolation: the phasic *pattern* of vasopressin secretion from a single cell has no meaning in itself—the pattern is lost in the averaging of activity of many cells. The patterning is, in the case of vasopressin, the mechanism whereby a population collectively can filter out transient inputs, and so it can only be understood by understanding the behavior of the whole population. The amount of vasopressin that is secreted from any single vasopressin cell is *not* linearly proportional to plasma sodium—far from it. However, the average secretion from the population *is*, because the population is very heterogeneous—some cells are more sensitive and some less sensitive than others. This heterogeneity gives the vasopressin response a wide dynamic range, much wider than that of any individual cell.

The flip-flopping of vasopressin cells is far from unique. In many neurons, including Purkinje cells of the cerebellum and pyramidal cells of the neocortex, neuronal activity can flip between stable states generated either by neuronal networks or through a variety of intrinsic mechanisms. As David McCormick put it in 2005, "Computation in the brain is not simply a matter of gathering influences from synaptic inputs and integrating these into a decision to spike. Spontaneous activity, generated both through intrinsic and network mechanisms, provides the context under which content is interpreted. Without context, all is lost."[20]

In the case of a vasopressin cell, how it responds to a given input depends on the state of spontaneous activity at the exact time the input arrives: this determines whether it will be inhibited, excited, or unaffected. The mean level of spontaneous activity also makes a difference: an excitatory stimulus that arrives when all vasopressin cells are relatively quiet will have a big effect, though it has none when the cells are all active. Thus slow factors that affect spontaneous activity can change how neurons respond to signals. Moreover, some slow signals can also affect the intrinsic properties of neurons, changing how they process information in many other ways: the properties that determine whether they fire in bursts, and in what type of bursts, can themselves be changed by peptide signals.

11 Vasopressin

The osmoreceptors, wherever they may be, do not accommodate during short-period exposure to a rise in the osmotic pressure of the carotid plasma produced by NaCl, doubling the period of exposure to the same rise causing the release of at least double the amount of antidiuretic substance. This lack of accommodation would be expected on the view that the osmoreceptors are continually engaged in controlling the antidiuretic function of the pituitary.

—Ernest Basil Verney (1894–1967)[1]

Our survival depends upon many things our brains and bodies do that we either scarcely notice or take for granted. The proper functioning of cells, organs, and tissues require a stable, safe, and sustaining environment. This requires, among many other things, that our bodies maintain a stable water content and a stable salt content. These are problematic because we constantly lose water and salts; we lose water through respiration; when we exercise we lose water and salt through perspiration; and we lose water and salt through urination. We gain water when we drink and salt when we eat, but access to drinking water may be erratic and dangerous, and for many animals, especially herbivores, getting enough sodium in their diet can be difficult, while some foods bring an excess of salt that must be excreted. A further problem is that dehydration is a *continuing* challenge; as noted by Verney, in the passage that introduces this chapter, the neurons that regulate our responses to dehydration cannot adapt as the neurons of the frog retina do, but must be able to maintain a steady response over many hours.

The hormone that was formerly called *antidiuretic hormone* and is now called vasopressin controls water loss from the body. After we drink a large amount of water, vasopressin secretion is suppressed, and we produce dilute urine to excrete the excess water. After a few hours without water, vasopressin secretion is increased again, and acts at V2 receptors on tubular epithelial cells in the renal collecting ducts.[2] This makes the epithelial cells more permeable to water: we now produce less urine (*antidiuresis*), and what we produce is more concentrated. How critical this is can be seen in pathological conditions where little vasopressin is produced, and an extreme case is that of the Brattleboro rat.[3]

In 1961, Dr. Henry Schroeder was raising a colony of rats for his laboratories in West Brattleboro, Vermont. In one cage, the bottle of drinking water was always quickly emptied, and this was noted by an alert assistant. The cage held a mother rat and seventeen pups, some of which were drinking copiously. Schroeder speculated that they might be affected by diabetes insipidus, a rare disease caused by the failure to produce vasopressin. This indeed seemed to be the case, because giving vasopressin to the affected rats prevented their excessive drinking, and Schroeder gave four of the rats to Heinz Valtin at Dartmouth Medical School, who bred them to raise the strain of "Brattleboro rats." These rats have a mutation in the vasopressin gene that prevents any vasopressin from being packaged into vesicles. Rats homozygous for this defect secrete no vasopressin and cannot produce concentrated urine: to retain a normal fluid balance they must drink almost their own body weight in water each day and must excrete a similar volume of dilute urine.

Familial diabetes insipidus is a rare condition in humans that arises from a similar mutation that prevents vasopressin from being packaged properly in neurosecretory vesicles. This disease often manifests suddenly in adult life, although the gene defect has been present since birth. It seems that the struggle to produce enough vasopressin to meet secretory demands causes overactivity of the neurons, and aberrant precursor molecules accumulate in the cells, affecting their normal function. As some cells become dysfunctional, the secretory demand falls increasingly on the remainder, leading them in turn to become dysfunctional, until, at a critical point, the system abruptly collapses.[4]

Although familial diabetes insipidus is rare, conditions of excessive vasopressin secretion are quite common. Sometimes this reflects the presence of a tumor, but in humans vasopressin secretion often increases with age. Excessive vasopressin secretion produces water retention and low plasma sodium levels, and in the elderly, these can cause mental confusion, and can affect gait. They also affect bone: bone is a major reservoir of sodium in the body, and with chronic sodium deficiency bones become increasingly fragile. Thus, in the elderly, disorders of vasopressin secretion can produce a vicious combination of effects that make falls more likely and more likely to result in fractures.[5]

The kidneys are extraordinarily sensitive to vasopressin. In humans, vasopressin normally circulates at about 1 pg/ml, and a tenfold increase will produce near maximal antidiuresis.[6] The main factor that controls its secretion is the plasma concentration of sodium: secretion is linearly proportional to this above an apparent threshold, a *set point*, which lies close to the normal sodium concentration.

Vasopressin is like the passenger who becomes more agitated the faster you drive, and unlike the passenger who is calm when you travel below the speed limit but who shrieks when you creep above it. The second passenger will be better at keeping your speed constant, so why does the vasopressin system respond proportionately to imbalance? Vasopressin cannot rectify an imbalance in plasma sodium; the only ways to achieve that are either by drinking water or by stimulating sodium excretion. All it can do is to slow the rate of progression of an imbalance, so why doesn't it do so as strongly as possible?

The pituitary stores a massive amount of vasopressin. In the rat, about 8,000 vasopressin cells project there, and each axon has about 2,000 swellings and nerve endings where vasopressin is stored in 7 billion large vesicles. Like all peptides, vasopressin is synthesized as part of a large precursor molecule, and in the cell bodies the precursor is packaged into the vesicles. These molecules are packed so densely that the core of each vesicle is virtually a solid crystal, leading these vesicles to be designated "large dense-core vesicles." Spikes that are propagated down the axons depolarize the swellings and nerve endings, causing some of these vesicles to be released. The secreted vasopressin enters the blood, which carries it throughout the body to bind to receptors on its target cells. The half-life of vasopressin in the blood is just a few minutes; what is secreted is what is needed now.

Abundant though these vesicles are, their rate of release is miserly. To maintain a normal plasma concentration (1–2 pg/ml), each cell must release just two or three vesicles each second. At this rate, the gland contains enough to last about 15 days, but with a tenfold increase in secretion it would take only about two days to empty the stores. By this time, vasopressin synthesis has been escalated and so secretion does not collapse, but the margin for error is narrow because it takes several hours to increase synthesis and to transport the hormone to its site of release. This might explain why the response of vasopressin is proportionate to imbalance: the stores are not so large that they can be casually frittered. Perhaps it is less important to keep plasma sodium constant than to keep it within safe limits. Given that we might find ourselves where water is unavailable, the vasopressin stores must keep plasma sodium within safe limits for as long as possible.

Vasopressin is secreted at a rate proportional to plasma sodium over about a tenfold range. A typical vasopressin cell is quiet at the normal sodium concentration. As this rises, its firing rate increases, a phasic pattern emerges and spiking activity is organized into bursts. As plasma sodium increases further, the bursts become more intense, but the bursts are still separated by long silences, and eventually the bursts give way to continuous fast spiking activity. However, there is considerable heterogeneity: even at high plasma sodium concentrations some cells are almost silent, and at low concentrations some are quite active.

The spike activity of each cell increases in proportion to plasma sodium, but its secretion does not. A vasopressin cell is like a sticky tap: you turn it and nothing comes out; turn it a bit more and you get a torrent. What is worse, the torrent subsides as the stores become exhausted—now the tap is fully open but just a few drips come out.

The issue is one of *stimulus-secretion coupling*:[7] while spikes are often thought of as the unit of information transfer in the nervous system, this is not true of any neuron. Neurons communicate by chemical messengers-whose release is coupled nonlinearly to spike activity. There are many nonlinearities: (1) The release of transmitter at an axon terminal is governed by the calcium that enters, and when spikes are clustered together, calcium transients can summate and potentiate release. (2) Only vesicles docked at the membrane are available for release, and spikes are less effective when this "readily releasable pool" is depleted. (3) The calcium entry triggered by each spike can depend on how fast a cell is firing; in vasopressin cells spikes broaden at higher firing rates, potentiating calcium entry. (4) Spikes may not invade all axon terminals; propagation can fail at branch points. When a spike is fired, potassium leaves the neuron and the extracellular concentration rises; with continued spike activity, this can depolarize the axon network so that spikes fail less often. (5) As calcium levels rise in the terminals this can impede further calcium entry and can hyperpolarize the terminals by activating potassium conductances. (6) What is secreted from vesicles can feed back onto the terminals, inhibiting or potentiating subsequent release.

The isolated posterior pituitary can be maintained in oxygenated medium for several hours, what it secretes can be measured every few seconds, and secretion can be stimulated by using electrical stimuli to trigger spikes in the axons that innervate the gland. By the use of this preparation, it became apparent that secretion depends on the pattern of stimulation, not just on the number of evoked spikes: 150 spikes at 50 Hz release about 100 times as much oxytocin as the same number of spikes at 1 Hz, and this *frequency-facilitation* of secretion is what makes the milk-ejection burst such a potent trigger of pulsatile oxytocin secretion during suckling.[8] For vasopressin, things are different: frequency facilitation saturates at about 13 Hz, and there is a converse process of secretory *fatigue*, so high rates of secretion cannot be maintained for long. The most efficient stimulation alternates activation at up to 13 Hz for about 20 seconds with silent periods of similar duration—the very pattern in which vasopressin cells fire when activated by dehydration.

This seems to be evidence of an "optimization principle." Vasopressin cells seem to fire phasically because that produces a given rate of secretion for the fewest spikes. From this we might infer that there is a penalty associated with too much excitation, and an obvious penalty is the risk of seizures. Whenever a spike is generated, the potassium that leaves a neuron will tend to excite neighboring neurons, so it is rapidly removed; glial cells, interleaved between neurons, clear it into the blood. If these mechanisms are overwhelmed, an uncontrolled spread of excitation can ensue that can kill neurons, a process called *excitotoxicity*. Dehydration requires a response that might need to be maintained for several days, requiring continuously high activity of the vasopressin cells, and so efficient stimulus-secretion coupling might be very important.

Thus we might conclude that vasopressin cells fire phasically because this is efficient for secretion. But what is most efficient for them is not most efficient for oxytocin cells, and other cells are different again. Phasic firing is efficient because the properties of the terminals have coevolved to make it efficient. Vasopressin cells are the way they are, but not how they *must* be; we need to look deeper.

At the normal plasma sodium concentration, a vasopressin cell fires at perhaps one or two spikes each second and secretes two or three vesicles each second. At any given terminal, just one vesicle is released every 20 minutes or so, one for every thousand spikes that invade the terminal. This is just as well: a terminal contains only a few hundred vesicles and would soon be emptied if the release rate were much higher.

As plasma sodium rises, spike activity may now comprise bursts at 6 spikes per second that last about 20 seconds. This activity is much more efficient: it now takes just 250 spikes to release a vesicle from each terminal. The cell must replace what has been secreted, but the resupply is delayed; more mRNA must be produced, and it takes about two hours for this to result in new hormone and longer still to transport the vesicles to the terminals.

If plasma sodium rises further, the bursts give way to continuous spiking at about 10 spikes per second. Now it now takes just 100 spikes to release a vesicle from each terminal, and this can't last long. The readily releasable pool is swiftly depleted and the rate of replenishment is too slow to keep up. Before long the reserve pool will run down.

Think of the horn on a vintage motor-car, with a trumpet and rubber bulb. Press the bulb hard enough, and it will hoot loudly. The horn has a narrow dynamic range: press too gently and no sound comes out; press very hard and the noise is little more than if you press it just enough. To hoot again, you must release the bulb to let it refill with air; if you keep pressing, the hoot will peter out. This is like what happens in vasopressin cells. The increased spike activity becomes more and more futile: continually active cells become incompetent at actually secreting anything.

So what rescues the vasopressin system to enable it to smoothly track plasma sodium, producing a proportionately graded secretion? Although each vasopressin cell increases its activity as plasma sodium increases, that activity is erratic and discontinuous. But the cells are heterogeneous: most are quiet at normal plasma sodium and some are very active. Some have long bursts, others short bursts, and some don't burst at all. Some have a high threshold to plasma sodium, others a low threshold.

This can all be simulated with a computational model.[9] We now understand the mechanisms that underlie phasic firing and can generate model cells that are indistinguishable from real cells in their spiking behavior. Knowing the variability of vasopressin cells, we can construct a population of model cells with the same variability. Knowing the properties of stimulus-secretion coupling, we can make each model cell "secrete" realistically. We can model how the readily releasable pools deplete and how they are refilled and how the reserves are replenished by increased synthesis, and can simulate how the secreted hormone is cleared from the bloodstream.

This model works beautifully—up to a point. As plasma sodium rises, the most sensitive cells lock on to the optimal phasic firing pattern and are responsible for most of the vasopressin secretion. As plasma sodium creeps up, more cells join in, but the cells that are most sensitive are also the first to be depleted and the first to be overactivated, firing hectically but pointlessly. Remember the penalty for overexcitation: damage and potentially death.

The vasopressin cells avoid this; they talk to each other, and take turns at carrying the secretory burden. To do this, each cell must know what the others are doing. They can't know how much vasopressin is being secreted because what is secreted into the blood can't reenter the brain. They don't communicate with each other by synapses, but they release vasopressin and other messengers from their dendrites.

The list of those messengers is long and growing.[10] Vesicles that contain vasopressin also contain other peptides, including dynorphin. The cells release adenosine, a by-product of their energy expenditure, and they synthesize nitric oxide. These signals work in different ways on the vasopressin cells and on afferent nerve endings, and have different functions. As described earlier, dynorphin acts back on the vasopressin cell that releases it to control its spiking activity: vasopressin, released from the dendrites in much greater amounts, has more widespread effects on the population of vasopressin cells.

Vasopressin cells release vasopressin from their dendrites and the dendrites express vasopressin receptors, so the vasopressin cells can know how much is being released. The vasopressin inhibits vasopressin cells, and the receptors to which it binds are internalized and disappear for a while, hence a cell that releases lots of vasopressin will become insensitive to it. However, the stores of vesicles in the dendrites become depleted just as the terminal stores do. As dendritic release subsides, receptors return to the cell surface and sensitivity to vasopressin recovers. Now, the cells that are depleted will be inhibited by vasopressin released from neighbors that are not depleted. Thus vasopressin cells alternate between high activity, when lots of vasopressin is available for release, and low activity when they are depleted. The quiet periods allow them to rebuild their stores and rejoin the game. By cooperative alternation of activity, the cells can share the burden of secretion equitably.

What we have seen does not sit comfortably with conventional views of the brain. Electrophysiology has been a powerful tool and its returns have permeated our thinking about how the brain works, but the resulting understanding is cursed by overinterpretation. Because spike activity appears to have so much capacity for encoding information, we are tempted to think it in fact conveys a huge volume of information. When we find, in the spike activity of some neuron or another, correlation between some facet and some particular external events, it is tempting to think this is the message the cell is conveying.

A message is only a message if it can be understood by its recipient. Neurons can't decipher long and complex sentences. They have a short attention span, are easily distracted, and much of the time they aren't even listening. A neuron that is quiet at a particular time is likely also to be less sensitive to an input; a cell that is very active might be saturated; and a cell recently activated might be refractory to further activation.

Spike activity is not the output of any neuron, only one of several means by which some of its chemical signals are generated. These signals are generated unreliably and erratically and are recognized imperfectly by their targets. The message carried by the spike activity of a vasopressin cell makes no sense when considered alone. The important signal is generated by a cacophony of noisy and messy cells, and the miracle that demands to be recognized is that this population response is indeed clean, refined, and fit for purpose.

There is no one "vasopressin cell": each is an individual. They are among the most completely characterized neurons in the brain. In my lab alone, my colleagues and I have recorded the activity of thousands of oxytocin and vasopressin cells in a myriad of experimental conditions. We have studied them in vivo and in vitro, in pregnancy and lactation, in response to osmotic stimulation and hypovolemia, and studied how they respond to afferent inputs from different brain sites, to blood-borne hormones, and to peptides and neurotransmitters in normal and pathological conditions. No two vasopressin cells are the same. They have things in common—features that, for me, distinguish them almost immediately not only from oxytocin cells but also from every other cell type I have studied. Each has its own bursting pattern and subtly different pattern of activity within bursts, and its own particular sensitivity to stimuli. Each has slightly different membrane properties, slightly different complements of receptors, a slightly different balance of afferent inputs.

This heterogeneity is not by design but by accident. The patterns of gene expression in any neuron are not rigidly fixed by genetic nature, they arise from the unique experience of each cell in its life from birth to adulthood. The innervation of each cell is not predetermined with precision. Axons that reach the supraoptic nucleus may be guided there by developmental cues, but which particular cells each axon contacts is an opportunistic accident. There are mistakes; developmental cues are imperfect and some axons get lost or misled and make inappropriate connections. The brain has to be robust against such imperfections; the cost of doing everything perfectly is too high.

Vasopressin cells are complex, but this does not make them clever, and the differences between cells certainly do not make each cell uniquely clever. I am not interested in the idea that the brain does clever things because it hosts 100 billion clever machines. The wonder is that it does clever things with machines that are messy, noisy, and imperfect.

Three other things should also be apparent. First, some things that we think of as easy may be far from trivial for the complex machines that fill our brain. Other things that we might think are difficult come easily to neurons. Our intuitions in this regard are unreliable. Second, neurons release many different messengers that act in different ways on different targets and at different temporal and spatial scales. What they release depends on what they synthesize and, at any given time, on how much is available to be released. Understanding synthesis is as important as understanding spike activity: for vasopressin cells, we can't make sense of their spike activity without understanding the interplay with synthesis.

Finally, we must consider what we mean when we say that the spiking activity of a neuron "encodes" information. We normally think of a code as something that conveys information from a sender to a recipient, and this requires that the recipient "understands" the code. But the spiking activity of every neuron seems to encode information in a slightly different way, a way that depends on that neuron's intrinsic properties. So what sense can a recipient make of the combined input from many neurons that all use different codes? It seems that what matters must be the "population code"—not the code that is used by single cells, but the average or aggregate signal from a population of neurons.

In a now classic paper, Shadlen and Newsome considered how information is communicated among neurons of the cortex—neurons that typically receive between 3,000 and 10,000 synaptic inputs. They argued that, although some neural structures in the brain may convey information in the timing of successive spikes, when many inputs converge on a neuron the information present in the precise timing of spikes is irretrievably lost, and only the information present in the average input rate can be used.[11] They concluded that "the search for information in temporal patterns, synchrony, and specially labeled spikes is unlikely to succeed" and that "the fundamental signaling units of cortex may be pools on the order of 100 neurons in size." The phasic firing of vasopressin cells is an extreme demonstration of the implausibility of spike patterning as a way of encoding usable information, but the key message—that the only *behaviorally relevant* information is that which is collectively encoded by the aggregate activity of a population—may be generally true.

12 Numbers

Creative work, in geology and anywhere else, is interaction and synthesis: half-baked ideas from a barroom, rocks in the field, chains of thought from lonely walks, numbers squeezed from rocks in a laboratory, numbers from a calculator riveted to a desk, fancy equipment usually malfunctioning on expensive ships, cheap equipment in the human cranium, arguments before a roadcut.
—Stephen Jay Gould (1941–2002)[1]

If you ask a neuroscientist what he or she is trying to do, the answer will probably be that they are trying to understand what some part of the brain does, or how it does it. This invites us to consider what we mean by "understanding." What kinds of explanation do we regard as adequate, and adequate for what, exactly? We may want a narrative, a story that makes sense. But we also want a story that is "true" to our knowledge and experience and that will still be true to knowledge and experience yet to come. Our stories must be consistent with all that we think we know and they must go further: they must be predictive, and by those predictions they must be tested.

I studied mathematics at university and decided that I wanted to use it to help understand the brain. I came to realize that we didn't then know enough biology to make the math useful and didn't have the mathematical tools either, and so became an experimental neuroscientist instead. Now we have a fuller understanding of how at least some bits of the brain work and have powerful mathematical and computational resources, and I came back, in some of what I do, to mathematics.[2]

By studying mathematics I had learned to think in abstract ways, to look for the "shape" of a possible understanding. Mathematics gives a palette of examples of how complex behaviors can arise from simple systems and a mental repertoire of concepts to understand them by. Models are, in the end, just analogies; analogies that help us construct a narrative understanding. They can take many forms, from the "stick and ball" model of DNA that Watson and Crick built, to Hodgkin and Huxley's mathematical model of the action potential that became a cornerstone of neuroscience. There are many kinds of mathematical model. Even unrealistic models can help to inspire hypotheses, and these can be better than the pedantically complex models that serve some as a kind of archive in which to deposit tedious facts. Modeling is also a test of the completeness of our understanding. If our story of how something works holds good, then we should be able to translate it into equations, tell it "in silicon," and show that our assumptions really do lead to the conclusions that we claim.

A good model should be no more complex than it need be: it must identify things that are critical and dispense with those that are not. Not everything matters: natural selection refines biological systems by discarding things that are harmful, but it doesn't hone everything to a miracle of economy. There are costs to getting rid of things and these can be high. For example, exactly where a receptor is expressed and in what quantity will be determined by elements in the promotor region of the receptor gene. A mutation in that region may induce a beneficial change or a maladaptive change in receptor expression in a particular part of the brain or body, but the same mutation might also induce neutral changes in other places. Receptors might be expressed pointlessly at sites where there is no endogenous ligand present, or at levels too low to be meaningful. Suppression of this occurrence might require further mutations, but the selection pressure for those will be weak and the cost of mechanisms to repress extraneous expression might be high.

It's cheaper to leave stains on the carpet than to buy a carpet cleaner: a stained carpet can still do a good job as a carpet. It's also cheaper to assemble a machine from things that are lying around than to fashion new components. Evolution modifies and adapts, it doesn't invent from nothing, so neutral mutations accumulate. If what they do replicates the function of something else, it might make the system more robust, but it's wrong to see them arising *because* they make the system robust; that's just what happens.

It's not the job of a model to reproduce the actual carpet with all its pattern and its stains, but to reproduce the carpet for its function as a carpet. Nevertheless, a model has to be anchored to reality. To make it meaningful—predictive and testable—it has to be anchored by numbers. Most biologists don't like numbers; they are the currency of experiments and observations but we don't like them. For the most part I've left numbers out in this book because nobody really likes them. When we build a model—no, when we build any kind of meaningful understanding—we must attend to numbers. They bring us down to earth, sometimes with a bump.

Let's look at a big number: 350,000,000,000. This is about the number of vasopressin vesicles in your posterior pituitary gland. How often does a vasopressin cell secrete one of these? Does it help to remind you that each cell has about 2,000 nerve endings from which vesicles can be secreted? Or to remind you that vesicles are secreted in response to spikes that invade these endings, and invite you to think of a cell firing, on average, between one and five spikes each second?

In 1976, John Morris set about counting vesicles.[3] He measured the volume of the posterior pituitary of the rat, and the volume occupied by each vesicle. Examining thin sections of the pituitary with an electron microscope, he measured about 1,000 vesicles to get an accurate average diameter, and counted more than 150,000 vesicles to estimate their density. He concluded that the gland contained about 14 billion vesicles. The oxytocin and vasopressin content had been measured extensively by bioassays and radioimmunoassays, and neurophysin, which is present in equimolar concentrations with the hormones, had also been measured by radioimmunoassay. These different measurements come to close agreement; rats of the size, sex, and strain used by Morris contain about 1 μg of oxytocin and about 1 μg of vasopressin in the pituitary. The molecular weights of oxytocin and vasopressin are both about 1,000, so, by Avogadro's number,[4] 1,000 g contains about 6×10^{23} hormone molecules, and 2 μg contains 12×10^{14} molecules. Therefore each vesicle contains about 83,000 molecules of either vasopressin or oxytocin.

With Jean Nordmann in Strasbourg, Morris checked this.[5] They homogenized posterior pituitary glands, centrifuged them, and separated the fractions to obtain a fraction enriched in vesicles. They mixed a sample of this fraction with latex particles of the same diameter as the vesicles, embedded it in resin, cut thin sections, and counted the latex particles and vesicles. From the ratio of these counts, they calculated the density of vesicles, and compared this with the hormone content as measured by radioimmunoassays and bioassays. They came to the same answer: each vesicle contained about 85,000 molecules.

The vesicles do not contain only vasopressin or oxytocin but the entire precursor molecule from which the hormones are cleaved, which has a molecular weight of about 23,000. An average vesicle has a volume that can theoretically contain *at most* 94,000 precursor molecules, little more than the apparent actual content. Thus the molecules of precursor in a vesicle are packed in a solid, "crystalloid" form.

John Morris went on to study exactly where the vesicles were located.[6] The axons that penetrate the gland have many "nerve endings" that look like synaptic endings, but they also have large swellings that contain more vesicles than the endings do. About half of the vesicles were in swellings, a quarter were in nerve endings, and the rest were in undilated axons. Working independently, Jean Nordmann came to similar numbers: he estimated that the gland contains about 17 billion vesicles, of which 59% were in swellings and 29% in endings.[7]

Morris showed that vesicles can be secreted from any of these compartments; the only thing that seems to matter is how many vesicles are close to the cell membrane. Newly synthesized vesicles go first to the endings, from which they can be secreted relatively easily. After a while, those that have not been secreted will be relegated to the swellings, to become part of a "reserve pool." Vesicles can be secreted from swellings, but the stimulation has to be more intense: after an acute stimulus, most of what is secreted comes from the endings, but after sustained stimulation the swellings also become depleted.[7] Vesicles that have not been secreted after a few days in the reserve pool will be moved to particularly large swellings called *Herring bodies*: these are graveyards, where vesicles that have passed their "release-by date" are dismantled by lysosomes.

So, the rat hypothalamus contains about 8,000 magnocellular oxytocin neurons and the same number of magnocellular vasopressin neurons. Each of these sends an axon to the pituitary that gives rise to about 400 swellings, each containing about 2,000 vesicles, and about 1,800 endings, each containing about 200 vesicles.

When vasopressin is secreted from the pituitary, it enters small blood vessels that drain into the jugular vein. In the rat, normal plasma concentrations are about 4 pg/ml, and the plasma volume is about 4 ml. Every two minutes, about half of the vasopressin is removed by the kidneys and liver, and the same amount must enter the blood as is removed. From these numbers alone, we can calculate that each vasopressin cell secretes a vesicle every ten seconds or so.[8]

In humans, we can ask what rate of supply of vasopressin is required for concentrating the urine. When a normal man drinks a large glass of water, his plasma sodium concentration falls, and he will secrete less vasopressin and produce dilute urine. If now you give him an infusion of vasopressin, he will produce less urine, and the urine will be more concentrated. The human kidney is *very* sensitive to vasopressin, and the rate required to produce "normal" urine is about 3 pg/min/kg. So, for an 80 kg man, the normal rate of vasopressin secretion is about 240 pg/min, secreted into about three liters of plasma. Every minute, about 10% is removed by the kidneys and liver, so, when the rate of supply equals the rate of degradation, the plasma concentration will be about 0.8 pg/ml, close to measured concentrations.[9] This is equivalent to 28,000 vesicles secreted every second. In humans, vasopressin cells are larger than in the rat, and there about 100,000 of them. In normal conditions, each of these must secrete about one vesicle every 4 seconds.

Across mammalian species, the size of the pituitary (and of the brain) is approximately proportional to the log of body mass (humans are an exception, but our large brain is due to a disproportionate enlargement of the cerebral cortex). This has consequences for endocrine systems, because blood volume is related to body mass not logarithmically but linearly. While the supply of vasopressin is 40 times greater in a man than a rat, the blood volume is *400 times greater*. This difference is partly offset by a longer half-life of vasopressin, but even so, vasopressin concentrations are lower in humans than in rats, and the kidney is more sensitive to vasopressin.

I concluded above that, in the rat, in normal conditions, each vasopressin cell secretes about one vesicle every 10 seconds. After a day without water, the plasma concentration has increased tenfold and the pituitary content of vasopressin has fallen by a quarter:[10] although synthesis is increased, the rate of supply always lags behind secretion. Now, about one vesicle is released every second from each cell: one vesicle every 30 minutes from each of 2,000 endings and swellings. After three days without water, the pituitary stores are nearly exhausted, and new vesicles are secreted almost as they arrive at the pituitary: the system is operating at its limits, with no remaining reserves.

In normal conditions, the vasopressin cells fire about one spike each second, and at any one ending, one vesicle is released for about every 20,000 spikes. After two days of dehydration, they are active in bursts at 6 to 8 spikes per second separated by silences. The secretion rate is ten times higher; now it might take "only" 5,000 spikes to release one vesicle.

These conclusions don't derive from any one study or any one method. The data on which they depend have been replicated in different species with different assays with diverse study designs, in many laboratories. At any one release site the release of vesicles is a very, very rare event.

It is inconceivable that such rare events are rigidly determined by spike activity; there must be a noisy probabilistic relationship between spike activity and these events. It is only the large numbers involved that give an illusion of determinacy—the many endings in a cell, the many cells in the population. If just one vesicle is released from each ending every 30 minutes, that is still 10,000 vesicles per second.

I have gone over these results carefully. The first reason is respect for the quality of the science. The data are not new or contentious: they came from rigorous systematic quantification, careful cross-checking for consistency with independent evidence, replication by independent workers, and replication with alternative methods. For this type of work the absence of a preformed hypothesis is a strength: there's less risk of confirmation bias, the malignancy that afflicts so much published work. Understanding how these data were derived makes it possible to assess sources of errors and miscalculations, and the layers of internal and external consistency give these particular numbers exceptional authority. The data are what they are and cannot be far wrong.

The second reason is that these measurements have important implications for our understanding of the brain, as explored in the next chapter.

13 Whispered Secrets and Public Announcements

It is impossible that the *whisper* of a faction should prevail against the voice of a nation.
—Lord John Russell (1792–1878)[1]

As explained in the last chapter, a single nerve ending in the pituitary gland typically contains about 200 peptide-containing vesicles. In conditions of intense physiological stimulation, about one of these vesicles is released every 30 minutes, during which time perhaps 6,000 spikes will have invaded that ending. These nerve endings have often been likened to synaptic endings in the brain, which often contain some large, peptide-containing dense-core vesicles as well as much more abundant small synaptic vesicles that contain classical neurotransmitters. Certainly they look similar, and the molecular mechanisms by which dense-core vesicles are released are very like those by which synaptic vesicles are released. For these reasons neuropeptides are described in many journal articles and textbooks as another class of neurotransmitters, as though, like neurotransmitters, they carry a message specifically from one neuron to another. This description is, to a first approximation, a lie.[2]

More than 100 different peptides are released in the brain. To glimpse the complexity, consider just one small nucleus in the hypothalamus, the arcuate nucleus. This is one of many nuclei in the hypothalamus—others mentioned in this book are the supraoptic, suprachiasmatic, paraventricular, periventricular, and ventromedial nuclei, the lateral hypothalamus, and the rostral hypothalamus—others include the dorsomedial, mammillary, and supramammillary nuclei and the preoptic area.

The arcuate nucleus is at the base of the hypothalamus adjacent to the median eminence. It hosts three populations of neuroendocrine neurons:

one secretes GHRH, which controls growth hormone secretion. Another secretes dopamine to regulate prolactin secretion; these fire in synchronous bursts every minute, and also express a peptide, met-enkephalin. Another population of dopamine neurons innervates the intermediate lobe of the pituitary to regulate α-MSH secretion into the blood. Other neurons control feeding: one population makes NPY and AgRP, both of which stimulate feeding, while another makes three peptides: α-MSH,which, in the brain, stimulates sexual behavior and inhibits feeding; the opioid peptide beta-endorphin; and CART (*cocaine- and amphetamine-related transcript*), which also inhibits feeding. Yet another neuronal population makes three more peptides kisspeptin, neurokinin B, and dynorphin, packaged in three different populations of vesicles. These regulate the pulsatile secretion of GnRH. Yet another makes somatostatin, and there is some evidence that another makes ghrelin. This exuberance is not unusual, and the list is far from exhaustive even for the arcuate nucleus. Most nuclei in the hypothalamus contain many clans of neurons that perform different physiological functions and express a diversity of peptides in addition to classical neurotransmitters. Many of these clans express several different peptides, in relative amounts that seem to vary from cell to cell. Just as all vasopressin cells express some oxytocin and all oxytocin cells express some vasopressin, it seems that all of the α-MSH cells make a small amount of NPY and all of the NPY neurons make a small amount of α-MSH. Neurons are messy things.

It is a widespread presumption that neuropeptides are an extended class of neurotransmitters: that they act at synapses, that their effects are tightly localized, and that their important effects, like those of classical neurotransmitters, are to alter the excitability of postsynaptic cells. This presumption, in all its facets, is wrong.

Vasopressin is the most abundant peptide in the brain. Most is made in the magnocellular neurons, but it is also made by neurons that regulate the secretion of ACTH from the anterior pituitary: *this* vasopressin is secreted together with another peptide, corticotropin-releasing hormone, into the blood vessels that supply the anterior pituitary. Vasopressin is also abundant in the suprachiasmatic nucleus, which controls daily rhythms of behavior, physiology, and hormone secretion, and these rhythms are entrained by light information that is carried by yet another population of vasopressin neurons, in the retina. Centrally projecting vasopressin neurons of the paraventricular nucleus regulate blood pressure and body temperature. Vasopressin neurons are also present in the olfactory bulb, anterior olfactory nucleus and piriform cortex, where they regulate social recognition by olfactory cues. In all of these neurons, vasopressin is packaged in large dense-core vesicles. But each of the nerve endings in the pituitary contains far more of these vesicles than has been seen in *any* synaptic ending in the brain. In the brain, no synapses contain 200 large dense-core vesicles or anything like that many; they contain a handful at most. Even if any synapse releases dense-core vesicles at the same rate as they are released at nerve endings in the pituitary, what sense can we make of a signal that is released on average, once for every 6,000 spikes?

Neuropeptides are fundamentally different from classical neurotransmitters. At a classical synapse, each spike typically releases (on average) one synaptic vesicle—often none, sometimes two or three; most synapses are not terribly reliable. One synaptic vesicle contains about 5,000 molecules of a neurotransmitter such as glutamate, and this is released into a narrow synaptic cleft, acts on receptors on the postsynaptic site, and is rapidly removed by transporters to be recycled. Everything is over in a few milliseconds. In the synaptic cleft, the concentration of neurotransmitter reaches very high levels, and the receptors at which it acts require these high concentrations. Peptide vesicles carry a much larger cargo (about 85,000 molecules), and their receptors are sensitive to concentrations a thousandfold lower than receptors for neurotransmitters. Peptides are broken down slowly, with half-lives that are generally a few minutes—at least 10,000 times longer than those of neurotransmitters.

We have to understand peptides as being released not from single neurons but from populations, and released in a coordinated way. Consider the magnocellular oxytocin neurons that generate milk-ejection bursts during suckling. The bursts occur every ten minutes or so for perhaps 12 hours each day, and are synchronized among 8,000 neurons. Each neuron has two dendrites: if a burst releases just 5 vesicles from each of these, then 80,000 will be released. Over a day, the average release rate from each cell would be just 1 vesicle per minute, and the total daily release would be less than a tenth of the total dendritic content.

What is released in the brain is degraded within brain tissue by aminopeptidases, and what survives arrives in the cerebrospinal fluid (CSF), from where it is cleared into the circulation. Concentrations of oxytocin are much higher there than in the blood, but only some of the oxytocin released in the brain reaches the CSF without being degraded. We can get closer to the true release rate by measuring neurophysin. Neurophysin is co-released with oxytocin in equimolar amounts, but is not significantly degraded in the brain, and for every molecule of oxytocin in CSF there are about 50 of the associated neurophysin. The rat cerebral ventricles contain about 7 pg of oxytocin—4 billion molecules—and about 200 billion molecules of neurophysin. Neurophysin is cleared from the CSF with a half-life of about 40 minutes, so we can calculate that, on average, about a million vesicles enter the CSF every 40 minutes—one every 20 seconds from each oxytocin neuron.

Clearly therefore, the notion that 10 vesicles are released from each oxytocin neuron during a milk-ejection burst is a conservative estimate. This, from 8,000 neurons, will deliver about 7 billion molecules into the hypothalamus, enough to raise the average concentration of oxytocin to 100 pM in a fluid volume of 0.1 ml. In the brain, the extracellular space is about a fifth of the tissue volume, so 0.1 ml is about the volume of extracellular fluid in a tissue volume of 0.5 ml. *This is close to the total tissue volume of the rat forebrain.*

Now, I am *not* saying that the oxytocin released in a milk-ejection burst floods the forebrain, only that enough is released to do so. Although it seems there is more than enough oxytocin in dendrites to deliver a widespread "hormonal" signal in the brain, we must consider what concentration is in fact effective at neurons, how far oxytocin released in the hypothalamus reaches, and where and how it is inactivated.

The first question, what concentration is effective, is surprisingly difficult to answer. Pharmacologists describe the actions of drugs by measuring how some response of a particular cell type varies with the applied concentration of that drug, and by measuring binding—how much of a drug is bound to receptors in a given tissue when different concentrations are applied. Generally, these measurements are reduced to single numbers, representing the concentration that will achieve a half-maximal effect (*EC50*), and the concentration that will achieve half-maximal binding (the *binding affinity*). These are only loosely related, because the effect achieved when a ligand binds to a receptor will depend on the properties of the cell and on the particular ligand. Peptides can have many different effects, and for each the EC50 may be different.

The human oxytocin receptor has an affinity for oxytocin of 0.28 nM,[3] meaning that oxytocin will bind about half of the oxytocin receptors that are exposed to this concentration. Affinity is a measure of probability—the probability that a receptor will be activated by a ligand at a given concentration. For any peptide, there are two ways of achieving a greater effect—you can either increase the concentration of the peptide, or increase the density of receptors: the effects are equivalent. Thus you can't predict what will be a physiologically effective concentration without considering both the density of receptor expression and the *efficacy* of the peptide—how many receptors must be activated to achieve a given effect.

This means that, while the concentrations of vasopressin and oxytocin in CSF are higher than are needed to activate peripheral tissues, we cannot be sure they are sufficient to activate neurons in the brain. Different cells in different brain regions might require much lower or much higher concentrations.

The second question, about how peptides disperse in the brain, is even more complicated. The extracellular fluid of the brain is not static, it is in constant motion. The traditional view has been that CSF is produced by the choroid plexuses in the lateral ventricles, and drains into the lymphatic space and subarachnoid space to be absorbed into the blood. More recent views suggest a much more complex picture, with continuous bidirectional fluid exchange at the blood-brain barrier. Movement of extracellular fluid is still more complex, and involves bulk flow alongside blood vessels and fiber tracts.[4] At present, we do not have a clear understanding of the direction and speed of flow of extracellular fluid in different brain regions.

The third question, about where and how peptides are inactivated also has no simple answer. Peptides are degraded by enzymes—some (like the opioid peptide β-endorphin, produced by some of the arcuate neurons) are apparently not degraded at all in the brain, while others have a half-life of just a few minutes. Oxytocin is degraded mainly by *placental leucine aminopeptidase*, also known as oxytocinase. Oxytocinase is expressed by neurons in a regionally specific manner and is mainly anchored to the membranes of those neurons. It seems inevitable that oxytocin released in a milk-ejection burst will flood the hypothalamus and adjacent regions, but exactly where it reaches is hard to predict; there's too much that we don't know.

Oxytocin neurons during a milk-ejection burst fire about 100 spikes in two seconds, and I have suggested that this releases just 10 vesicles of oxytocin in the brain. A classical neuron firing this way would release about 100 synaptic vesicles from each of 10,000 axon terminals; because many terminals end on the same cells, perhaps 1,000 neurons will be affected. But the oxytocin cell is not alone, and when all are active together, their collective signal will act throughout large areas of the brain. The anatomical connectivity becomes irrelevant: all that counts is where the receptors are expressed and in what amounts. Whereas neurotransmitters are whispered secrets that pass from one neuron to another at a very specific time and place, peptides are public announcements, broadcast to whole populations.

What is true of oxytocin and vasopressin, that their effects are not localized to synapses, has to be true of all peptides in the brain. Any neuron that contains any large dense-core vesicles—and most neurons in the brain do—can release them only very infrequently, and will release these vesicles not consistently from a few sites, but sparsely from many different sites—mainly from the varicosities or swellings that stud most axons in the brain. Even if one vesicle might affect just a few neurons close to the site at which it is released, the different sites at which these vesicles are released will affect different groups of neurons and do so erratically. But when many neurons of a peptide-secreting population are activated together, the net effect will be a local, hormone-like release of peptide that indiscriminately affects all neurons that express the necessary receptors in the particular brain region that they innervate, with long-lasting effects.

Peptides operate on multiple scales: they have feedback effects on the cells of origin that modulate activity patterning, and local effects on neighboring cells to coordinate the behavior of a population; and the hormone-like release of peptides from cell populations can have organizational effects on distant targets. It's a mode of communication quite different from neurotransmitter release. Oxytocin, as we have seen, by its priming actions, can affect how oxytocin cells communicate with each other. How common such priming actions are we don't know. But all peptides can affect gene expression and can alter the behavior of neurons by changing what receptors they express and what they secrete. These actions of peptides together underlie what we might see as a *reprogramming* of communication in the brain.

14 Plasticity

Whether 'tis nobler in the mind to suffer
The slings and arrows of outrageous fortune,
Or to take Arms against a Sea of troubles,
And by opposing end them:
—William Shakespeare, *Hamlet*

About twenty years ago I was invited to be an instructor at a training school for new researchers in endocrinology. The school was funded by the pharmaceutical company Ferring, and was held on the island of Föhr in the North Sea, at the home of Fredrik Paulsen, the founder of Ferring. The high points of the week, for me, were two after-dinner talks given by Fredrik, one on the history of Ferring, and another on *language*. Föhr is a small island with just a few thousand inhabitants, but seven languages are spoken there, including one, *Fering*, which is spoken only on Föhr. Fredrik's theme was that every language is a unique repository of knowledge and understanding.

Föhr had been a whaling community, and in the early race to develop treatments for the diseases of hormone deficiency, whales played a small but important part. As might be expected, the pituitary of the whale is large and a rich source of hormones very like those of humans. Moreover, the anterior pituitary is separated from the posterior pituitary by a bony plate, making it easy to separate the tissues without cross-contamination, so Fredrik recruited whalers to salvage the pituitaries for research. In Fering, the pituitary is known by an idiom that translates as "the button on the brain." Whalers had long recognized this conspicuous gland and had noticed that, when a whale had fought particularly long and hard, the pituitary would be engorged with blood. This explains another idiom—"the button on his brain has burst," applied to someone overcome by stress. Fredrik argued that knowledge of the pituitary's involvement in stress was embedded in the language, long predating our scientific awareness.

Anyone who attempts to track the historical development of physiological concepts is likely to become frustrated at their imprecision and instability. Terms that appear to define specific biological entities, like genes, neurotransmitters, and hormones, have accrued multiple different and sometimes incompatible meanings that depend on the context in which they are used. Today, in some contexts, a gene is the sequence of DNA that encodes a specific protein; in others, it is the longer unit that spans the exons that encode the protein sequence but also includes the introns that separate the exons and which themselves include sites that regulate the expression of the gene product; in yet others, it is a longer and ill-defined sequence that includes all regulatory sites. Sometimes, such as when we talk of the gene for a certain trait that differs between individuals, we mean a variant version of some site of regulation of a gene. Sometimes we talk loosely about genes as though they were synonymous with specific traits, as in the "genes for hair color." Sometimes, even more loosely, we talk of genes "for" diseases like cystic fibrosis or Huntington's disease, as though their purpose is to give us these diseases.

These confusions are as nothing to those entailed by terms that have connotations from their use in normal language. As I write, I have just returned from a meeting with collaborators in a multidisciplinary project to study the determinants of food choice. We spent our last few hours together reflecting on how we use the term *hunger*, an important reflection if mutual understanding is to be more than an illusion. We variously use "hunger" to refer to: a state where there is not enough available energy to meet current and expected demands; the signals that inform the brain of that state; the effects of those signals on specific brain areas; our conscious awareness of that state; and the motivation to eat regardless of how that motivation arises. We also use diverse "operational definitions" of hunger such as self-reported scales of perceived hunger and measures of the willingness to work to receive a food reward. That these are all different things we must keep in mind whenever we talk. "Hunger" is a ghost in the machine of the brain that assumes different and often intangible forms, and it will not be easily expunged.

"Stress" is even more problematic. Our current use of the term was coined by Hans Selye about eighty years ago. Selye had shown that diverse physical and emotional threats all produced the same three pathological outcomes: enlarged adrenal glands, stomach ulcerations, and lymph tissue shrinkage, and he chose the word *stress* to describe this triad, defining it as "the non-specific response of the body to any demand for change." Defining stress in this way forced him to introduce a new term, *stressor*, for things that produce stress, but keeping the terminology tight and consistent was not easy. Paul Rosch, one of Hans Selye's collaborators, later reflected: "Even Selye had difficulties, and in helping to prepare the First Annual Report on Stress in 1951, I included the comments of one critic, who, using verbatim citations from Selye's own writings, concluded that 'Stress, in addition to being itself, was also the cause of itself, and the result of itself.'"[1]

Neuroendocrinologists try to avoid such confusions by using operational definitions tied to the secretory activity of the adrenal cortex; thus a *stressor* is any stimulus that results in increased production of glucocorticoid hormones—mainly cortisol in humans, corticosterone in rodents. These steroid hormones are not stored in the adrenal cortex but must be produced on demand, and their production is regulated mainly by ACTH from the pituitary. Both the production and the secretion of ACTH is regulated by factors released, as we now know, by neurons in the paraventricular nucleus (figure 14.1).

The triad of hypothalamus, pituitary, and adrenal cortex is commonly known as the *HPA axis*. This triad does not capture the full range of physiological responses to stressors: the "fight or flight" responses to acute threats that involve increased heart rate and blood pressure, increased glucose production by the liver, and secretion of adrenaline from the adrenal medulla are regulated by other paraventricular neurons via a pathway that engages the autonomic nervous system. Nor is the HPA axis the only endocrine system engaged: the secretion of many hormones is affected by stressors. However, it is the HPA axis that produces stress as defined by Selye.

Glucocorticoids act on many organ systems. In the liver, fat, and muscle they mobilize energy to meet current or anticipated demand, and they also *redirect* energy. For example, they suppress inflammation, which is a response to injury that, however useful, might impair the ability to escape from a continuing threat. Glucocorticoids engage long-lasting adaptations,

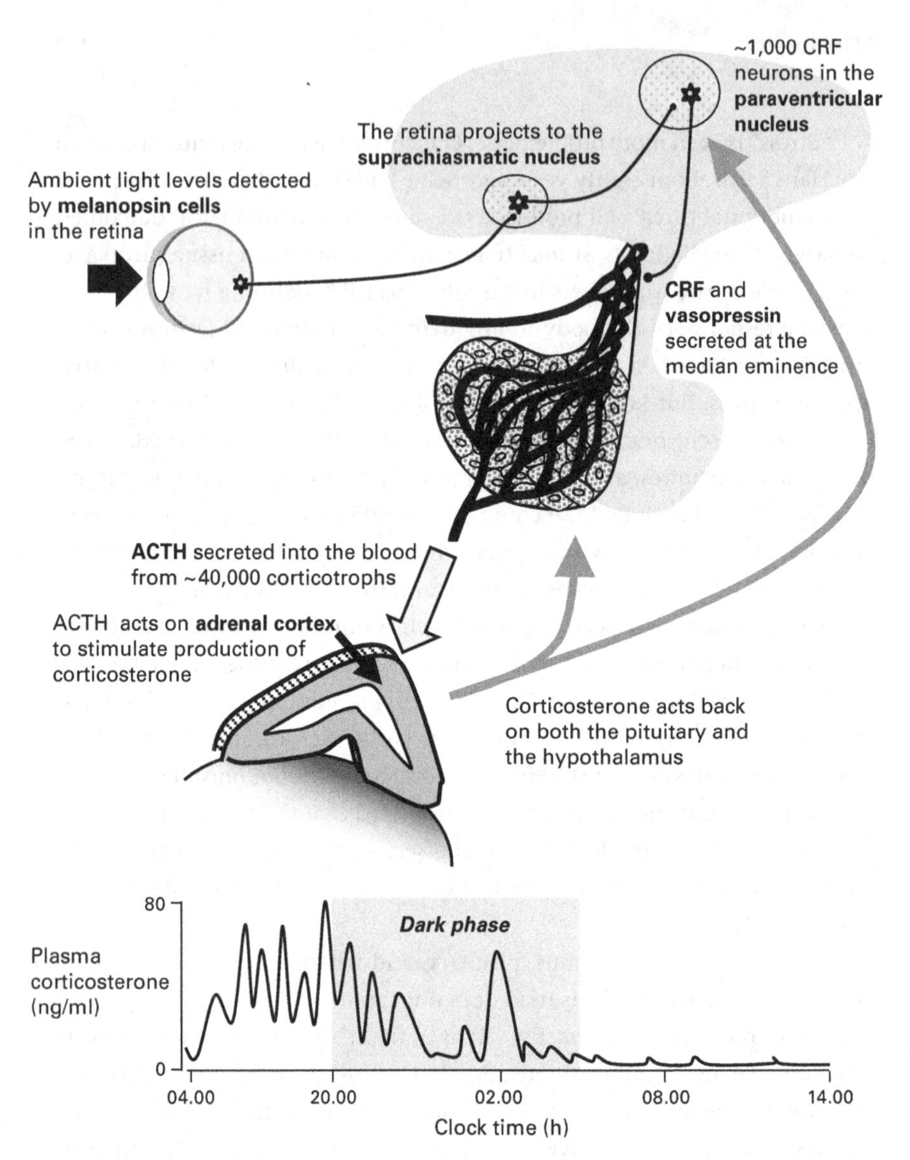

Figure 14.1

Amplification and feedback in the stress axis. In the rat, CRF (and vasopressin) is produced in about 1,000 neuroendocrine neurons of the paraventricular nucleus. These neurons project to the median eminence, where they release brief pulses of CRF into portal blood vessels that carry the CRF to the anterior pituitary. There, CRF stimulates the corticotrophs to secrete pulses of ACTH, which reaches levels of up to 1 ng/ml in the blood. ACTH acts on about 400,000 cells of the adrenal cortex to stimulate the production of corticosterone, which reaches levels of 80 ng/ml and more in plasma. This acts back on both the pituitary and the brain. The secretion of corticosterone follows a diurnal rhythm—rats are nocturnal, and highest levels occur at about the time they become active. This daily rhythm is governed by light signals, detected by the retina and communicated to the paraventricular nucleus via the suprachiasmatic nucleus. Illustration modified from Spiga F et al.[2]

and can do so because they affect the expression of many different genes in different tissues. Glucocorticoids are lipid soluble, and so enter cells freely; inside a cell, if a molecule of glucocorticoid encounters a glucocorticoid receptor the two molecules will bind together, and this complex is ferried into the cell nucleus where it binds to "glucocorticoid response elements" in the DNA. There are such elements in the regulatory regions of many genes.

Glucocorticoids act not only on peripheral organs, but also on the brain and pituitary. They signal through at least two receptor subtypes, the *mineralocorticoid receptor* and the *glucocorticoid receptor*: both are expressed in the brain, especially in the hypothalamus but also in the hippocampus, which is involved in learning and memory. This should be no surprise: if an innocuous event is a reliable harbinger of a threat that we can avert by an appropriate behavioral or physiological response, we need to learn this and remember it. A stressor given shortly after learning will generally strengthen memory, especially for stimuli that evoke emotional responses, whereas stressors before learning are distractors. Mineralocorticoid and glucocorticoid receptors are "nuclear" receptors, affecting gene expression, but glucocorticoids can also exert rapid effects on neuronal excitability by effects on membrane receptors.

Glucocorticoids are needed to assure energy availability even in the absence of stress. They are secreted in a daily rhythm, mobilizing energy in anticipation of the normal cycle of energy use. This rhythm is modulated by signals from the periphery that indicate how much energy is available; in rats, sucrose intake will dampen glucocorticoid production, while low levels of free fatty acids, sensed by neurons in the paraventricular nucleus, enhance it. Superimposed on this daily rhythm is a pulsatile pattern of secretion; every hour or so a pulse of ACTH triggers a pulse of glucocorticoid production. Surprisingly, this does not depend on a pulsatile input from the hypothalamus, as revealed in a mathematical model by Stafford Lightman, John Terry, and Jamie Walker.[2] Because ACTH triggers the *production* of glucocorticoids, which takes time, there is a delay between ACTH secretion and the glucocorticoid production that it evokes, and glucocorticoids then act back on the pituitary to suppress the secretion of ACTH. Given a constant hypothalamic drive to ACTH secretion, this cycle will result in a pulsatile pattern of secretion very like that normally observed.

Any novel stimulus is a potential threat, and will evoke a sharp peak of ACTH secretion and glucocorticoid production. We might see this as a signal that puts brain networks "on alert." The signal goes everywhere, but only recently activated pathways are affected. In the hippocampus, circuits that contain glucocorticoid receptors receive information about diverse sensory stimuli. If a stimulus proves innocuous, then it may have no enduring consequences. However, if it has an injurious sequel, then a large and prolonged rise in glucocorticoid concentration will act on those pathways that have been put on alert to consolidate the memory of the stimulus and encode its linkage with an adverse consequence.

This account just gives the shape of a possible explanation consistent with present knowledge and understanding. If neuroscientists sometimes talk as though they know all the answers, it's a hubris that is often easily punctured. Our theories are partial and provisional. Paul Rosch, reflecting on Hans Selye's contributions, wrote, "His real legacy can be summed up by what he often reminded me, namely, that theories do not have to be correct. Only facts do. Some theories are of value because of their *heuristic value, in that they encourage others to discover new facts, that then lead to better theories.*"[1]

By looking at the effects of glucocorticoids on the brain, we can see that an indiscriminate hormonal signal, accessing all regions of the brain, can nevertheless have actions that are specific, sophisticated, and enduring. The specificity is only partly because only some neurons have glucocorticoid receptors: the effects of glucocorticoids also depend on *context*, including the context of recent events, and these effects can have prolonged consequences for the behavior of neurons.

The survival value of associating a stimulus with a threat is obvious, and the association must be learned from a single experience and remembered for a lifetime. In travels to many parts of the world I have eaten and usually enjoyed many different foods. One thwarted me. *Lutefisk*, long part of Swedish cuisine, is dried cod that has been soaked in a lye solution for several days to rehydrate it before being rinsed with cold water to remove the lye; it is then boiled or baked. My only encounter with it was at a restaurant in Stockholm. Before a spoon touched my mouth I felt my face turn white, broke into a sweat and started to gag. Something had triggered in me the unconscious memory of something encountered, probably in my childhood, of which I have no conscious memory. Memories don't have to involve any conscious elements or reasoning processes, and it's often better that they don't.

If the HPA axis is continually activated, as in many disease conditions that are accompanied by chronic pain or inflammation, then the prolonged production of glucocorticoids can have serious pathological consequences. So, while it is important that we learn that some stimuli are threats, we must also find ways of coping with stressors that we cannot avoid, but which turn out not to be unduly harmful.

Rats, when first shown a dark tube, will readily enter it—they like dark places. But if the tube is then closed so they can't escape, they become anxious and produce large amounts of corticosterone. If a rat is restrained in this way for an hour each day, the corticosteroid response habituates; each day, it is less than on the previous day. It seems that the rat learns that the experience, though initially unsettling, is not harmful, and you might imagine that this is a reasoned understanding, that the rat learns that nothing untoward ensues and hence does not activate corticosterone production. This is sort of true, but it happens in an interesting way. To understand what happens, we have to understand something of how the paraventricular nucleus regulates ACTH secretion.

After Geoffrey Harris had proposed that secretion from the anterior pituitary was regulated by hypothalamic factors, the first confirmation of their existence came from evidence that extracts of the hypothalamus contained a factor that potently stimulated ACTH secretion. However, this advance was followed by frustration; it seemed impossible to disentangle the effects of any "true" corticotropin-releasing factor (CRF) from those of vasopressin. Vasopressin was always present in tissues that had CRF activity: it could stimulate ACTH secretion, but did not seem potent enough to be *the* CRF. It did not seem credible that vasopressin could regulate both ACTH secretion and antidiuresis: the levels in plasma seemed too low, and only some stimuli that activated ACTH secretion also activated vasopressin secretion. Studies in the Brattleboro rat, which lacks vasopressin, produced conflicting results: some claimed that the HPA response to stress was normal, but others that it was attenuated. If vasopressin was *the* CRF, then the Brattleboro rat should have profound deficits in ACTH secretion and corticosterone production—which it does not.

In the late 1970s and early 1980s, several findings emerged in quick succession. It was found that not all of the vasopressin neurons in the paraventricular nucleus projected to the posterior pituitary: some projected to the median eminence, so there was a *second* vasopressin system, one that might be implicated specifically in ACTH secretion. After removal of the adrenal glands, the vasopressin content of the paraventricular nucleus massively increased, suggesting that its release was regulated by hormonal feedback from the adrenal glands, as would be expected of a hypothalamic system of regulation. So, if vasopressin was not *the* CRF then surely it had to be a major part of it. There seemed to be three possibilities: that vasopressin was involved in regulating the secretion of a CRF; that a CRF and vasopressin were independently regulated releasing factors for ACTH; and that vasopressin acted as a releasing factor in concert with a CRF.

Phil Lowry and Glenda Gillies had advocated the theory that vasopressin was the main factor regulating ACTH secretion, and when they found that other hypothalamic factors with "weak, labile CRF activity" could potentiate vasopressin-induced ACTH release, they proposed, in a letter to *Nature,* that "CRF is vasopressin modulated by other hypothalamic factor(s) released into the hypothalamo-hypophyseal portal system."[3]

Just two years later, Wylie Vale and his colleagues at the Scripps Institute succeeded in purifying, from extracts of sheep hypothalami, a peptide with extremely high potency for stimulating the secretion of ACTH from culture.[4] There was little doubt that they had found the physiological releasing factor, and it was not vasopressin. Soon, the term CRH (*corticotropin-releasing hormone*) superseded the term CRF as it became generally accepted that the peptide identified by Vale was indeed a hormone that regulated ACTH secretion.

Gillies and Lowry did not let the matter rest. The following year, again in *Nature,* they published a letter with the title "Corticotropin releasing activity of the new CRF is potentiated several times by vasopressin."[5] They had studied ACTH secretion from isolated pituitary cells; they confirmed that the new CRF was more potent at secreting ACTH than vasopressin, but found a synergistic interaction between this CRF and vasopressin. *Synergy* is a term that is often used loosely when an effect exerted by two factors in combination is greater than the effect of either alone. Usually such effects are *additive,* rather than synergistic. True synergy can be seen when two drugs that are both applied at maximally effective concentrations evoke a response that is much greater than the sum of their separate effects. Synergy in this sense cannot be explained by two drugs acting on the same receptors, but implies an interaction between the intracellular signaling pathways that the two drugs engage.

Then, two years later, in 1984, Wylie Vale and colleagues reported that they had found the neurons that make CRH, as expected, in the paraventricular nucleus. Between 1% and 2% of them also contained vasopressin. But when the adrenal glands were removed from rats, about 70% of the CRH neurons contained vasopressin. They concluded that "there is a state-dependent plasticity in the expression of biologically active peptides by individual neuroendocrine neurons."[6]

It was well known that the brain is plastic—that experience and physiological state can alter the connectivity of neurons—but the idea that what messengers these neurons use might also change was novel. Here was a neuronal system that used a potent ACTH secretagogue, CRH; now, in addition it seemed to be making another, vasopressin, which alone was less potent than CRH but which *in combination* with CRH was extremely potent.

In Bristol, Ming Ma and Stafford Lightman began studying how the expression of the genes for vasopressin and CRH in rats changed in response to repeated episodes of restraint stress.[7] The first episode evoked a large rise in corticosterone levels, and it stimulated the synthesis of both vasopressin and CRH, as measured by levels of heteronuclear RNA (the immediate copy of the coding region of the DNA before it is processed into mRNA) in the paraventricular nucleus. This is to be expected: whenever a peptide is released, more must be made to replenish the releasable pools. When the stressor was given daily, the corticosterone response declined progressively, and after two weeks it was completely absent. The CRH response declined similarly. So far, the account is as might be expected if the stimulus ceases to become stressful—if the rat learns that it is innocuous.

However, whereas the response of the CRH gene to the stressor decreased with repetition, that of the vasopressin gene to the stressor *increased*. The neurons in the paraventricular nucleus were still reacting to the stimulus—but differently. If with each episode of a stressor the neurons increase the synthesis of vasopressin but not that of CRH, then progressively their stores of releasable peptide will include more vasopressin and less CRH, and what were CRH-secreting neurons will become vasopressin-secreting neurons.

To see what this might mean we have also to know that although rats exposed to a repeated stressor will habituate to it, they will still react to a new and different stressor—indeed, repeatedly stressed rats show a *heightened* response to some different stressors.

To make sense of this, we must recognize that different stressors signal to the paraventricular nucleus by different pathways, and do not target exactly the same neurons. So, consider one type of stressor, let us call it stressor A, that activates a subset of CRH neurons—subset A. With repeated exposure to stressor A, these become vasopressin-secreting neurons, and because vasopressin is less potent than CRH, they release less ACTH. A different type of stressor, stressor B, may activate another subset B that are still CRH neurons, so the ACTH response to stressor B is intact although that to stressor A has diminished. The interesting case is stressor C, which activates some neurons from subset A and some from subset B. This will evoke a mixture of vasopressin and CRH release, and so can give a supranormal response.

The phenotypic plasticity of CRH neurons can no longer be regarded as exceptional. Across the hypothalamus, more than a hundred different peptides are expressed in different neuronal subpopulations along with several hundred specific peptide receptors. Many neurons produce several peptides, in a bewildering variety of combinations, and this Kandinsky-like canvas is in constant turmoil, as the expression of each of these genes waxes and wanes with the multifarious rewards and insults of daily life.

One of the most startling changes has been described in the neurons that regulate prolactin secretion. When the pituitary is separated from the hypothalamus, the blood levels of most of its hormones disappear, but prolactin levels increase, because its secretion is governed by a factor that inhibits spontaneous secretion. That factor is dopamine, released from the arcuate nucleus. Dopamine is better known as a neurotransmitter in neuronal circuits within the brain, especially in the "reward circuits" of the brain and in the circuits involved in fine motor control that are affected in Parkinson's disease; there is no single "dopamine system" but many systems doing very different things. Even in the arcuate nucleus there are at least two populations of dopamine neurons; one regulates prolactin secretion, while another innervates the intermediate pituitary and regulates secretion of α-MSH from melanotrophs into the blood.

Electrophysiological studies of dopamine neurons in the arcuate nucleus indicate that they, rather like vasopressin cells, alternate every 20 seconds or so between an up state during which they fire at about four spikes per second and a quiescent down state. But unlike vasopressin cells they fire in synchrony, releasing dopamine in frequent pulses. Prolactin has a long half-life in blood (about 20 minutes), so the rapidly pulsating dopamine release is clearly not there to drive a slowly pulsing prolactin secretion. However, it is possible to study the activity of the prolactin gene in real time in individual lactotrophs using a "reporter gene" in which the expression of a light-emitting protein (luciferase or green fluorescent protein) is governed by the same DNA sequence that regulates prolactin expression.[8,9] When this gene is introduced into a lactotroph, the level of prolactin expression can be inferred from the intensity of light emitted from the cell. These studies have shown that transcription of the prolactin gene is highly pulsatile. In cell cultures, different lactotrophs display very different and asynchronous fluctuations, but within explanted pituitary glands they are less heterogeneous and are bound together by intercellular signals. It is not yet possible to study transcription in real time in lactotrophs in the living animal, but it seems likely that a common rhythmic drive from pulsatile dopamine release coordinates transcription of prolactin among the whole population of lactotrophs.

The functions of prolactin are many and diverse. In fish and amphibians, prolactin is an osmoregulatory hormone, regulating the permeability of the skin to water and salt, and it also regulates reproduction. In some birds, prolactin promotes fluid production in the crop sac to feed fledglings—a function analogous to milk production in mammals. In mammals, prolactin is involved in many aspects of reproduction, and also in metabolic regulation through receptors expressed on adipose tissue, liver, pancreas, and the brain. Prolactin promotes appetite, and this might underlie the weight gain in pregnancy that anticipates the nutritional demands of suckling young. It also causes infertility, by suppressing LH secretion, probably through its actions on GnRH neurons, and is the main reason why women who are exclusively breastfeeding an infant seldom become pregnant.

Prolactin also suppresses libido in both males and females and enhances parental behaviors. When male talapoin monkeys are housed together, they sort themselves into a dominance hierarchy. The most aggressive and sexually dominant males have high levels of testosterone and low levels of prolactin, while the most subservient have high levels of prolactin and low levels of testosterone. These endocrine differences are a *consequence* of the emergence of a hierarchy, not the cause of it, and they have important consequences. Subordinate individuals withdraw from competition for mates, which might be seen as a way of avoiding injuries that would be incurred by fighting; since dominant males are also those most at risk of injury, we might see subordinate individuals as biding their time.[10]

However, the best-known role of prolactin is in lactation, when it stimulates the production of milk. Prolactin secretion is stimulated by suckling, and evidence indicates that this must involve an inhibition of dopamine release. Prolactin secretion is regulated in part by a negative-feedback loop—as levels in the blood rise, some is transported into the brain, where it activates the dopamine neurons to inhibit further secretion. This transport mechanism is unchanged in lactation, so if the dopamine neurons are quiescent in lactation despite continued high levels of prolactin, then either something is inhibiting them strongly, or else something is stopping them being activated by prolactin.

Dave Grattan and his colleagues in Otago and Montpellier set about testing these hypotheses, and elegantly refuted both. They found that the electrical properties of the dopamine neurons were unchanged in lactation, and so were their responses to prolactin. However, in one important detail things were different: in lactation, they no longer produced dopamine. Instead, they were making met-enkephalin,[11] by which they signal not to the pituitary but to other neurons in the arcuate nucleus: met-enkephalin, among other things, is a potent stimulator of appetite.

To put all this in a broad context, in response to signals from the periphery, including hormonal signals from the adrenal glands and the gonads, some neurons in the hypothalamus don't merely change their electrical activity, they also change the messengers that they produce and who they talk to. The pattern in which they release their messengers not only affects the electrical and secretory activity of their targets, it can also affect the expression of genes in those targets. We are glimpsing a system that is not like a giant supercomputer executing some vast and sophisticated program, but like an ecology of multiple small computers that are constantly reprogramming themselves or being reprogrammed by external events.

15 Rhythms

I arose (joyous and full of love) at cockcrow. How good seemed everything at that hour, my darling! When I opened my window I could see the sun shining, and hear the birds singing, and smell the air laden with scents of spring. In short, all nature was awaking to life again. Everything was in consonance with my mood; everything seemed fair and spring-like.

—Fyodor Dostoyevsky (1821–1881), *Poor Folk*, chapter 1

In all animals, the transition between night and day engages a host of physiological and behavioral rhythms. These rhythms depend not on the classical photoreceptors—the rods and cones of the retina—but on retinal ganglion cells that express melanopsin, a different photopigment than those present in rods and cones, and which was first discovered in photosensitive skin cells of the African clawed toad. The retinal ganglion cells that express melanopsin are not involved in analyzing visual images, but in detecting the ambient light level, and they project directly to the suprachiasmatic nucleus of the hypothalamus. This nucleus is often described as the brain's "master clock": its 20,000 neurons (in the rat) govern daily (*circadian*) rhythms of behavior and hormone secretion.[1] Whether a master clock is the right analogy is debatable: the suprachiasmatic nucleus wakes us up of a morning, much as a cockerel might do, but nobody so far seems to have wanted to describe it as a "master cock."

However, the suprachiasmatic nucleus does not simply respond to dawn's early light. Even when kept in constant darkness, mammals still show rhythms of behavior and of hormone secretion (notably the secretion of growth hormone, ACTH, and glucocorticoids) that consistently follow a cycle of about 24 hours. These rhythms, of sleep and wakefulness, of drinking and eating behavior, of exercise and body temperature and hormone secretion, define periods of subjective "day" and "night" in the absence of any light cues—or indeed any other environmental cues: they are generated intrinsically in the brain. These rhythms can be entrained by a short light pulse given just once a day—the start of the subjective day can be "reset" by a brief light cue. However, if the suprachiasmatic nucleus is lesioned, then this circadian rhythmicity is lost, and it is lost even in animals kept in normal lighting conditions.

In 1988, Ralph and Menaker reported that while studying the circadian rhythms of hamsters kept in constant darkness they had noticed that one male hamster had an unusually short rhythm, of about 22 hours, not the normal 24 hours.[2] They bred this male with normal females, and by crossing the offspring produced a line of hamsters—*tau mutant hamsters*—that all had the same short rhythms. This showed that a mutation in a single gene—a gene then unknown—could alter the period of intrinsically generated circadian rhythms.

In the 1980s, several groups reported that in rats and hamsters in which circadian rhythms had been abolished by lesions of the suprachiasmatic nucleus, the rhythms could be restored by transplanting the suprachiasmatic nucleus from a newborn animal into the hypothalamus. These studies attracted great excitement, for not only did they show that the suprachiasmatic nucleus was the generator of these circadian rhythms, they also seemed to indicate the viability of "brain transplants," as the transplanted tissue appeared to be functionally viable and to have integrated itself appropriately into the host brain. However, when the brains of these animals were studied microscopically, little anatomical integration was evident.

Then, in 1996, Rae Silver and colleagues showed conclusively that such transplants could restore circadian rhythms without any anatomical integration of the tissue at all.[3] They showed that transplants could restore circadian rhythms in hamsters bearing lesions of the suprachiasmatic nucleus even if the transplanted tissue was wholly enclosed in a semipermeable capsule. Moreover, if the tissue that they grafted came from a tau mutant hamster, then the recipient of the graft showed circadian rhythms with a period of 22 hours—the period of the tau mutant hamster. These were stunning findings. They showed that the tau mutation affected the intrinsic properties of the suprachiasmatic nucleus, and also that this nucleus could signal to the rest of the brain by a diffusible signal without any anatomical connections.

Knowing that a single gene mutation could have a subtle yet telling effect on circadian rhythms fueled the hunt for "clock genes" in the suprachiasmatic nucleus. Fruit flies (*Drosophila*) also show circadian rhythms of activity, in which oscillations in the level of expression of the gene *period* (*per*) were known to be important. In 1997, Tei and colleagues reported that homologs of this gene were present in humans and mice, and that the mouse version of the gene was expressed in the suprachiasmatic nucleus, where the levels of expression displayed autonomous circadian oscillation.[4] By 2001, it was clear that at least eight clock genes were somehow involved: three *period* genes (*per1*, *per2*, and *per3*); two *cryptochrome* genes (*Cry1* and *Cry2*); *double-time* (*dbt*, the gene involved in the tau mutation); and two "master genes," CLOCK and BMAL1.[5]

To describe a complex molecular mechanism very simply, the ability of clock cells to generate a rhythm with a period of about 24 hours comes from a set of "clock genes" that they express. In clock cells, a first set of genes produce factors that activate a second set of genes which then produce factors that act back to repress expression of the first set. There can be several hours between activation of the expression of a gene and the production of its protein product. These delays produce a *transcriptional/post-translational delayed feedback loop*—a regular cycle of gene expression, the period of which depends on particular features of the genes involved. The products of these clock genes are all transcription factors, and these affect many other genes in the cells, giving rise to circadian rhythms in receptor expression and peptide expression.

Many cells express clock genes and can generate their own 24-hour rhythms that can be entrained by various external events. For example, daily arrival of food entrains a food-entrained oscillator that generates anticipatory responses to food arrival; this does not involve the suprachiasmatic nucleus but engages populations of neurons at other hypothalamic and extrahypothalamic sites. There are also clocks in peripheral organs, including in the pancreas, and in adipocytes. What distinguishes the suprachiasmatic nucleus is that it sends a signal to the rest of the hypothalamus to coordinate circadian rhythms of behavior, physiology, and hormone secretion—rhythms that in turn entrain peripherally generated circadian rhythms.

An important part of that signal is carried by vasopressin. The suprachiasmatic nucleus contains about 5,000 neurons that produce vasopressin, and these neurons project to many parts of the hypothalamus. One site that they project to is the OVLT, and in mice, vasopressin released there at the end of the night stimulates thirst—promoting a surge of water intake just before these nocturnal animals go to sleep.[6]

In rats kept in constant darkness, the expression of vasopressin in the suprachiasmatic nucleus oscillates in a 24-hour cycle, and when the rats are given periodic flashes of light, the cycle will become locked to those light signals. The rhythm of gene expression is entrained to the light-dark cycle by retinal afferents (from the melanopsin-containing retinal ganglion cells). In the suprachiasmatic nucleus, these inputs terminate on two clans that make VIP (vasoactive intestinal peptide) and GRP (gastrin-releasing peptide), which act on the suprachiasmatic neurons that make vasopressin.

Mike Ludwig had been studying vasopressin neurons in a line of transgenic rats generated by Yoichi Ueta and colleagues in Kitakyushu.[7] Ueta had engineered these rats to express a fluorescent protein in the vasopressin neurons, and Ludwig had already noticed that the protein was expressed not only in the expected sites—the supraoptic and paraventricular nucleus and the suprachiasmatic nucleus, but also in some populations of neurons outside the hypothalamus. One of these populations was in the olfactory bulb, and Ludwig, along with myself and other colleagues, showed that these neurons were involved in the recognition of social odors. Intrigued by this, Ludwig went on to look closely at other sensory systems, and found that some retinal ganglion cells also expressed vasopressin. He assembled a consortium of colleagues to study them, including Valery Grinevich in Tubingen, who engineered the cells to express a fluorescent protein that enabled the axons of the cells to be traced, and who showed that they projected to the suprachiasmatic nucleus. Another collaborator was Javier Stern in Georgia, who recorded from these cells in vitro and showed that they were intrinsically light sensitive. What exactly was the role of vasopressin in *this* system?[8]

Ludwig argued that it was probably important that we *don't* rapidly adjust our circadian rhythms whenever we experience a change in lighting. Animals are sensitive to a short period of light at the end of the dark period—they will rapidly reset their rhythms to this, but they are not sensitive to a short exposure of light that is given early in the dark period. We wake up early in the first case, but quickly fall back to sleep in the second. Pursuing this idea, we found that in rats, light given at the end of the dark phase consistently evoked measurable vasopressin release in the suprachiasmatic nucleus of rats, but light given at the beginning of the dark phase did not.

The retinal ganglion cells that project to the suprachiasmatic nucleus use glutamate as a neurotransmitter, secreted from small synaptic vesicles. Because glutamate can be rapidly recycled, this signaling is constantly available. However, vasopressin is not contained in the same vesicles as glutamate; it is packaged in separate, large vesicles that are synthesized at the cell body and transported along the axons, and these vesicles cannot be recycled as small synaptic vesicles are. Although vasopressin synthesis will be increased to replenish what has been released from the terminals, it takes several hours to produce more vasopressin and to transport the new vesicles to the terminals. At the terminals in the suprachiasmatic nucleus, the availability of peptide for release must be subject to a diurnal cycle of depletion and replenishment. Thus, we argued, when light comes early in the night, little vasopressin is released, but when it comes late in the night the vasopressin stores are full and waiting.

In annelids, vasotocin cells contain a light-sensitive protein, an *opsin*, suggesting that the peptide is secreted directly in response to changes in light conditions.[9] The vasopressin cells of the mammalian suprachiasmatic nucleus are not intrinsically sensitive to light, but vasopressin is also expressed in retinal cells that project to the suprachiasmatic nucleus, and those cells express the light-sensitive protein melanopsin. It seems that the vasopressin cells in the hypothalamus have lost the intrinsic photosensitivity of their evolutionary ancestors, but those in the suprachiasmatic nucleus have remained associated with light-responsive functions, while those of the retina have retained intrinsic photosensitivity.

An important function of the suprachiasmatic nucleus is to control the *pineal gland*, an endocrine gland buried in the center of the human brain, but located on the dorsal surface of the rat brain between the cerebellum and cortex. The pineal gland produces melatonin (from the Greek for "tonic of darkness") during the hours of darkness in response to signals that reach it from the suprachiasmatic nucleus by a convoluted, multisynaptic pathway, and melatonin in turn acts back on the suprachiasmatic nucleus and at other sites in the brain via specific G protein–coupled receptors. In seasonal animals like sheep and hamsters, the melatonin signal regulates reproduction as well as other seasonally regulated physiological processes:

the melatonin signal encodes *day length,* and, in different species, reproduction is triggered by the seasonal shortening or lengthening of days. Humans are not seasonal animals, but melatonin is a very effective treatment for jet lag, probably because, by its actions on the suprachiasmatic nucleus, it can promote the resetting of circadian rhythms.[10]

Jet lag is an irritating encumbrance, but disorders of circadian rhythms have much more serious consequences in people. They are associated with metabolic abnormalities, heart disease, reduced immunity, increased stress, and abnormal cognition and mood states. Russell Foster and Leon Kreitzman eloquently summarized the importance of these rhythms:[11]

> Until we turned our nights into days and began to travel in aircraft across multiple time zones, we were largely unaware that we possess a "day within" driven by an internal body clock. Yet the striking impairment of our abilities in the early hours of the morning soon reminds us that we are slaves to our biology. Our ability to perform mathematical calculations or other intellectual tasks between 04.00 and 06.00 h is worse than if we had consumed several shots of whisky and would be classified as legally drunk. Biological clocks drive or alter our sleep patterns, alertness, mood, physical strength, blood pressure and every other aspect of our physiology and behaviour.

Indeed. And they do so by hormonal messengers within the brain, by the diffusible signals, including vasopressin, that are released from the suprachiasmatic nucleus, and by melatonin from the pineal.

16 Obesity

Some hae meat and cannae eat
Some would eat that want it
But we hae meat and we can eat
Sae let the Lord be thankit.
—Robert Burns (1759–1796), "The Selkirk Grace"

At the tail of the eighteenth century, when Rabbie Burns wrote "The Selkirk Grace," food was not to be taken for granted. Even in wealthy Edinburgh where Burns lived there were frequent riots in protest at the price of grain. At the University of Edinburgh, where I work, the Anatomical Museum displays the skeleton of a leader of those riots: "Bowed Joseph," barely four feet tall, had the curved spine and deformed limbs characteristic of rickets, once a common disease of undernutrition, but he had a powerful voice, and charisma. As told by Sir Walter Scott, Joseph could summon a mob of up to ten thousand "town rats" by the beat of his drum, "all alike ready to execute his commands, or to disperse at his bidding."[1] In 1780, returning drunk from Leith Races, he fell to his death from a stagecoach.

In the early nineteenth century, most people in Western Europe could afford only the cheapest foods: cereals, root vegetables, and pulses. Livestock products provided less than 15% of calorie intake, and little sugar, green vegetables, or fresh fruit was eaten. Through the nineteenth and twentieth centuries, the productivity of agriculture increased and drove down the real price of food. At the same time, with the spread of industrialization and trade, real incomes grew. The cost of moving food from rural areas into cities also fell, and when refrigeration was introduced in the 1870s, cheaper foods could be imported from distant agricultural economies. At first, people ate more bread and potatoes, but increasingly they ate more sugar, oils and fats, fruit and vegetables, much more meat and dairy products, and fewer starchy staples. In Northern Europe, the proportion of calories derived from livestock rose from less than 15% in the early nineteenth century to more than 30% in the 1960s, and this pattern was repeated across the world as countries developed.[2]

As countries get richer and the cost of food falls, and as life expectancy increases and reproduction rates fall, there will be more obesity:[3] obesity develops slowly, so its prevalence increases with age. As we age, we all produce less growth hormone, and in consequence we lose muscle mass and increase fat mass. When everyone has enough to eat, more are bound to become obese; those predisposed to obesity become victims of enough food to realize that predisposition.

Obesity is not an aesthetic judgment; worldwide; back in 2002, still rising rapidly, it was the sixth most important risk factor contributing to the overall disease burden. In developed countries, it was the fifth leading cause of loss of healthy life, behind tobacco, high blood pressure, alcohol, and high cholesterol. In the United Kingdom, obesity contributes to about 30,000 deaths every year, and about ten times that many in the United States, where it is now the main preventable cause of illness and premature death.[4]

Between 1971 and 2000, the prevalence of obesity in the United States increased from 14.5% to 30.9% of adults. This accompanied an increase in energy intake: according to food surveys, women increased their calorie consumption by 22% (from 1,542 to 1,877 calories per day) but men by only 7% (from 2,450 to 2,618 kilocalories). This mainly reflected increased consumption of carbohydrates, especially pasta, breads, and cereal-based snacks. The percentage of calories from fat decreased while protein consumption remained about the same.[5,6]

The cause of obesity might seem obvious. Although the world population is rising, food production is rising faster, and more people have the opportunity to become obese. We eat more energy-dense foods, and are less physically active due to the increasingly sedentary nature of work and increasing urbanization. Where our fathers toiled and sweated, we tweet and snack; where they trekked and climbed, we use cars and escalators.

This might be partly true, but it's dangerous to mistake a plausible explanation for a valid one. Scientists are professional skeptics. Skepticism isn't about making fun of nonsense like homeopathy—that's the job of every thinking person. It is about questioning not things we can see to be nonsense but things we take for granted. If people eat too much and take no exercise they are likely to become obese, but are we really eating more and exercising less? And if we are, is this really why more of us are obese?

The message from the United States seemed to be that obesity was due to overconsumption of carbohydrates, but most countries have seen an increase in obesity despite different consumption patterns. In Canada, the prevalence of obesity almost tripled between 1985 and 2000,[7] and there, energy intake increased by 18% between 1991 and 2002, but, unlike in the US, it was fat consumption that increased most.[8] In Australia, in the twenty years to 2000, the daily average calorie intake grew by only 87 calories yet obesity rates doubled.[9] In the UK, even while obesity rates rose, calorie intake declined, as did the percentage of fat in the diet[10,11] (figure 16.1).

So, depending on which data you choose to use, you can tell any story you like. The problem might be that these data are not perfect, especially data about diets that depend on people answering questionnaires. Those who respond to food surveys may not always confess their taste for "junk food," and there seems to be consistent underreporting of intake. Similar problems affect all surveys that rely on people being willing to disclose things they would rather deny to themselves. Nevertheless, it seems unlikely that the reported decline in food consumption hides an increase that is masked by increasing underreporting.

If we can't rely on people to tell the truth and we can't keep them in laboratories for years while we control their diets, what can we do? We have data on total food production. In the United States, the total food supply in 2000 provided 3,800 calories per person per day, 500 more than in 1970. But of those calories, the US Department of Agriculture estimated that about 1,100 were lost to waste, spoilage, and other causes, including cooking.[12] That's a big uncertainty, and if we waste more food now than we used to then the data will be very misleading. Comparing countries, we find little relationship between obesity and food supply. Those with low food production have low rates of obesity, but if we compare only industrialized countries there's a rather weak correlation. It is hard to be sure that we are eating more now than people did fifty years ago or that people are less active now, and hard to see any one globally applicable explanation of the rising obesity rates. There are differences between men and women, rich and poor, city dwellers and country folk, and these differences vary from region to region.

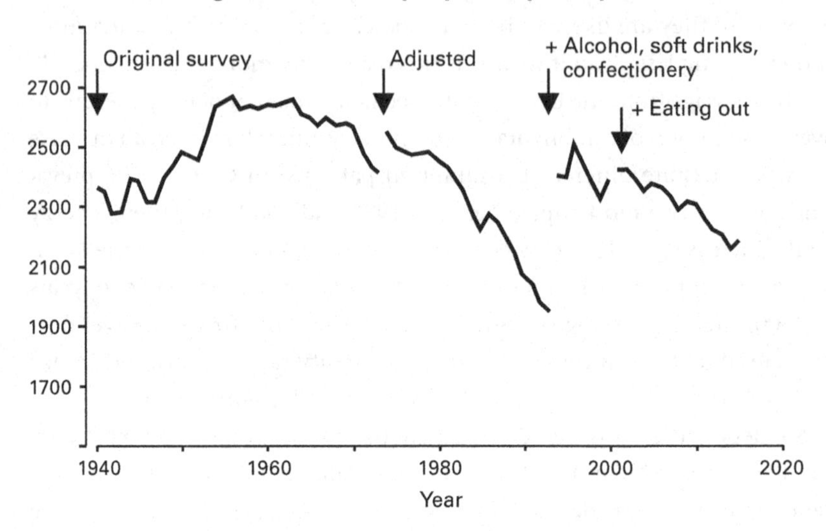

Figure 16.1
Trends in food intake. Data collected by an annual survey of food purchases in the United Kingdom. From 1940 to 1973, the data did not include alcoholic drinks, soft drinks, confectionery, or eating out. From 1973 to 1991, the reported intakes were adjusted. From 1992 onward the data included alcoholic drinks, soft drinks, and confectionery, and from 1994 they also included eating out. The data (https://www.gov.uk/government/collections/family-food-statistics/) were published on March 9, 2017, by the Department for Environment, Food, and Rural Affairs in the UK. Data were collected annually for a sample of households using self-reported diaries of all purchases over a two-week period. Where possible, quantities were recorded in the diaries, but otherwise they were estimated. Energy and nutrient intakes were calculated using standard nutrient composition data for each of 500 food types.

Obesity is perhaps best defined as an excessive enlargement of the body's adipose tissue. We might think of fat as an ugly nuisance, but it is a very important organ: it gives heat insulation and it stores energy for times of shortage; it allows us to eat intermittently, sometimes to indulge our appetite, sometimes to skip a meal for other, competing rewards. In white adipose tissue, specialized cells called adipocytes store fat in semiliquid form. This fat can be converted into energy in a way that is regulated by hormones.

Obesity is conventionally defined as a body mass index (BMI) of more than 30. BMI is a crude index, calculated as the ratio between body weight (in kilograms) and the square of height (in meters), but family doctors and outpatient clinics routinely monitor BMI in large numbers of patients, making data easy to collect frequently and on a large scale. For adults, a BMI of less than 18.5 marks you as underweight, above 25 as overweight, and above 30 as obese. For any one person, this may be misleading—athletes with a large muscle mass can have a high BMI despite a very low fat mass. Men with a "healthy" BMI typically have 15% to17% body fat, while women have between 18% and 22%. Typically, elite male athletes have just 6% to 12% body fat, and some have much less.

Most people have about 30 billion adipocytes, and these expand or contract according to the supply of energy. This storage system is flexible, but it can be overwhelmed, and different people have different capacities to store fat safely. When the capacity to store fat safely is exceeded, lipid can accumulate in the heart, skeletal muscle, pancreas, liver, and kidney. This increases the risks of many diseases, including heart failure and type 2 diabetes.

How fat you are depends mostly on your genes, genes that determine your metabolic rate and general body composition;[13] how fat your parents are is a good predictor of how fat you will be. The obesity epidemic has been blamed on lifestyle changes—the "obesogenic environment"—but exactly who becomes obese is largely determined by genetics. This confuses many people—if body weight is governed by our genes, how can we be getting fatter when they haven't changed? Understanding this requires understanding how genes interact with the environment.

Heritability is defined technically as the proportion of total phenotypic variability caused by genetic variance in a population. In this case, we are concerned with understanding how much of the variability in body weight in a population is due to the genetic differences between individuals. This can be addressed by looking at family pedigrees—how similar closely related people are in BMI—and especially by studying identical twins and nonidentical twins. The presumption is that twins are reared in the same environment; identical twins are usually much more similar in BMI than nonidentical twins who have some differences in their genes.

Such studies all show that human obesity has a strong heritable component.[14] Exactly how strong the link is between BMI and genes depends on sex, on the environment, and on when in life BMI is measured. In children, the link is strong: about 80% of variation in BMI appears to be due to heritable differences, with the family environment having surprisingly little effect. In a few individuals, obesity can be attributed to a defect in a single gene, but these cases of monogenetic obesity are very rare. Generally, the variability of body weight reflects differences in many genes, each with small effects on different physiological processes, including metabolic regulation, appetite, food preference, and fat storage.

Counterintuitively, the apparent heritability of BMI has increased as the obesity epidemic has unfolded. When food is in limited supply, individual differences in BMI mainly reflect environmental factors—variations in access to food. When food is cheap and abundant, genetic differences between individuals reveal themselves, and, because our genes encode a predisposition to weight gain that depends on the environment, these differences are magnified.

Why so many appear to be predisposed to gain weight has long been the subject of speculation about evolutionary mechanisms. The "thrifty phenotype" hypothesis proposed that obesogenic energy-efficient genes favoring fat storage emerged by natural selection in populations repeatedly exposed to famine.[15] By contrast, the "drifty phenotype" hypothesis proposed that when early hominids began to make tools, use fire, and band together they became less subject to predation, so genes that favored a lean phenotype became less subject to selection pressure.[16] This relaxation of selection pressure would result in the persistence of minor mutations, producing a diversity of human phenotypes including some prone to obesity. A third view acknowledges that genetic susceptibility to obesity is not the same across ethnic groups.[17] Its adherents propose that descendants of early humans who remained in Africa or migrated to tropical or subtropical environments maintained heat adaptation genes, while the descendants of those who migrated to colder regions acquired genes for cold adaptation, equipping them with a higher metabolic rate and greater resistance to obesity. Perhaps all are true, to some extent for some populations, perhaps none are: it matters *that* we differ more than *why* we differ.

The heritable differences in BMI have many components: differences in how much we eat depend on differences in appetite, but the amount of weight gain that results from overfeeding also has a strong genetic component, as does the impact of specific diets and that of exercise. However, the search for the genes responsible has (so far) found associations that account for only about a fifth of the variation between individuals.[18] This has focused attention on other heritable mechanisms, and particularly on the consequences of events in uterine and early postnatal life. Stress and poor nutrition during gestation and in early life have lifelong "programming" effects on physiology. If you were born with a low body weight, either because your mother was stressed during pregnancy or because of inadequate nutrition, you are more likely to become obese. Apparently, your physiology develops in anticipation of a world where food is scarce and must be consumed whenever possible.

This hypothesis was proposed to explain the findings of a now famous study. Toward the end of the Second World War, a blockade by the German army led to a famine in the west of the Netherlands. The Hunger Winter of 1944–45 killed more than 20,000 people. In 1976, a study of 300,000 recruits in the Dutch army showed that the sons of women who had become pregnant toward the end of the famine were more likely to be obese than men born before or after. The authors proposed that early nutritional deprivation had affected the development of the hypothalamic regions that regulate food intake and growth.[19] Subsequent studies in animals showed that the offspring of mothers who had been underfed while pregnant grew up to be hyperphagic: they ate bigger meals and preferred high-fat food. These changes were accompanied by changes in the hypothalamus, especially in the expression levels of neuropeptides that regulate food intake.[20,21]

Environmental challenges, however severe, cannot change our genes: the DNA sequence of an individual is fixed. However, how genes are expressed can be influenced by *epigenetic* modifications—gene sequences can be "silenced" by processes such as DNA methylation and histone modification. The children conceived toward the end of the Hunger Winter showed differences in the prevalence of methyl groups on the gene that encodes IGF2, a hormone that regulates fetal growth. Children whose mothers had gone hungry in early pregnancy had fewer methyl tags than siblings born earlier or later, and similar effects were found in mice whose mothers had been fed a low-fat diet while pregnant. These "programming effects" can have a big influence on disease susceptibility in later life, affecting the risks for heart disease, obesity, and diabetes.

The father's experiences also matter. Male rats on a high-fat diet become obese and develop impaired glucose tolerance and insulin sensitivity—signs of diabetes.[22] Surprisingly, their daughters are also affected: even on a normal diet they have impaired glucose tolerance and insulin sensitivity.

Thus, parents "experiences can affect their children's health. There is also evidence that some of these effects can be passed from generation to generation. One theory is that programming "adapts" a newborn to the world it is born into, by modifying its physiology as appropriate for an environment that is placid or stressful, lush or barren.

How fat you are also depends on your gut microbiota.[23] We are not alone, even in our own bodies: each of us shares our body with tens of trillions of microorganisms in our gut. These live on the nutrients we ingest, and their composition varies according to our diet and antibiotic exposure. There is a huge diversity of species, and they have many beneficial effects—they help in digesting nutrients and protect against some pathogens—but they have also been associated with inflammatory bowel diseases, irritable bowel syndrome, colorectal cancer, allergic disease, and obesity.

Identical (monozygotic) twins generally have a similar body weight and a similar gut microbiome (the composition of the whole population of microorganisms: which species are present, and in what proportions). But some identical twins differ substantially in body weight, and in 2014 a study published in *Science* suggested that such discordance might be due to a discordance in their gut microbiome.[24] That the leaner of a pair of twins has a different gut microbiome than an obese brother or sister is not surprising, and it might reflect no more than that some microorganisms thrive better than others in the guts of obese individuals. But this study took samples of the microbiota and introduced them into the guts of germ-free mice. Remarkably, mice given samples from the obese twins gained, on average, about 10% in fat mass, while mice given samples from the lean twins maintained a normal weight. In humans, fecal-matter transplants to alter gut flora are an effective treatment for recurrent, refractory *Clostridium dificile* infection, and some reports suggest that such transplants might also be an effective treatment for some cases of diabetes and obesity.[25,26]

How many calories a person needs depends on many things. An individual's energy expenditure can be measured by giving a dose of "doubly-labeled water," enriched with the isotopes ^{2}H and ^{18}O. Oxygen leaves the body either as exhaled carbon dioxide or in water (in urine, sweat, and breath), whereas hydrogen can leave the body only in water. So, by comparing the ratios of these isotopes in blood, urine, or saliva it is possible to calculate how much oxygen has been converted to carbon dioxide. This is a measure of the average metabolic rate, and is equivalent to daily energy requirements. People vary in their energy requirements according to their age, gender, and level of physical activity, but also according to their different genetic constitutions and their physiological status—pregnant and lactating women need much higher energy intakes.

We use energy in all our physical activity—climbing stairs, walking, even fidgeting. We use energy in eating and digesting food, and for raw foods this can be a considerable proportion of their energy content. We use energy in maintaining our body temperature, even if most of us who live in cold climates have centrally heated homes, sleep in warm beds, and stay mainly indoors. *Nonexercise activity thermogenesis* represents these common daily activities, and can result in up to 2,000 calories of expenditure per day above the basal metabolic rate.[27–30]

Are people less active today? We don't have objective data from fifty years ago. Fewer people have manual jobs, people walk less, and the most popular leisure pursuits are screen time and computer games. But more people consciously exercise, and more cycle. More than 45 million Americans belong to health clubs, up from 23 million in 1993. Does it make a difference how much we exercise? It takes a lot of effort to burn off excess weight, but we don't even lose the weight that we expect to for the exercise that we do take. In studies lasting more than 25 weeks the average weight loss is only about a third of that predicted from caloric expenditure. The Dose Response to Exercise in Women study examined the benefits of regular exercise for overweight postmenopausal women with elevated blood pressure.[31] Three groups of women took supervised exercise for six months, at 50%, 100%, and 150% of recommended intensity, and a control group had no supervised exercise. More than 400 women took part, and most, even in the control group, lost weight. However, the women who exercised hardest did not lose more than control subjects, and about a quarter of them gained weight. The difference between actual and predicted weight loss in response to exercise has been termed "compensation."

If we increase our food intake, exactly how much weight we put on varies from one person to the next, but typically we defend our body weight by increasing energy expenditure after eating more than usual. Conversely, if we exercise intensely, we compensate by reducing our energy expenditure between bouts of exercise or by eating more.[32] That's nothing new; as a child I was sent out on a Sunday morning for a long walk or to play football to build up an appetite before dinner.

There is little value in general recommendations about calorie intake; a recommended value that is enough for all will be too much for some and not enough for others. Health Canada produces detailed recommendations by weight, gender, and activity level.[33] They estimate that a woman aged 31 to 50 needs between 1,800 and 2,250 calories each day depending on her level of physical activity, while men of that age need between 2,350 and 2,900 per day. Remember that, in the United States, the estimated food supply in 2000 provided 3,800 calories per person per day, of which about 1,100 were wasted. This leaves 2,700 calories per day—not very different from Health Canada's advice on the intake required for a healthy body weight. The image of the obese individual as scoffing beefburgers and doughnuts washed down with sugary drinks is misleading. There are some like this, just as some are emaciated by self-denial, or by diets for which our digestive systems are poorly adapted. But the diets of most overweight people differ little from those of most people of normal weight.[34]

The physicians who must deal with the problems arising from obesity have little to offer. People on a strict calorie-controlled diet lose weight, but when the control stops, most regain the weight that they lost and often gain more. Exercise is hard for the obese, hard on the knees and joints, humiliating in public, and depressingly ineffective.

Obesity is not a lifestyle choice, nor a moral failing or a weakness of will, but a multifactorial disease, a disease that is often a dysfunction in the hypothalamus. This can reflect a dysfunctional responsiveness to hormonal and neural signals from peripheral tissues, dysfunctional production of those signals, or dysfunctions in the signaling within the hypothalamus from diverse populations of neuropeptide-secreting neurons that regulate appetite, food choice, energy expenditure, and glucose homeostasis.

It doesn't do much good to berate the obese for their lack of willpower. There's something offensive about the disdain of the affluent lean for their overweight neighbors, a disdain sometimes expressed by politicians for whom words come to mouth faster than facts to brain. Who knows what stress your neighbors encounter in their daily lives, the legacies of their genes or their birth environment, and what struggles they endure to restrain their weight?

Living in Edinburgh, you don't escape Robert Burns—the passionate poet of the common man, songster and seducer, full of wit and anger. His "Address to the Rigidly Righteous" is an attack on conventional, unthinking morality and the dogmatic beliefs of those who believe they can easily identify causes of human behavior.

Who made the heart, 'tis He alone
Decidedly can try us;
He knows each chord, its various tone,
Each spring, its various bias:
Then at the balance let's be mute,
We never can adjust it;
What's done we partly may compute,
But know not what's resisted.

17 The Empty Medicine Cabinet

Despite enormous investment from pharmaceutical companies and an international focus on research into obesity, today we have no drugs approved as safe and effective that have more than a small and transient effect on body weight.[1] Several have been withdrawn because of concerns about safety, and many others never reached clinical use because of side effects revealed in animal tests or early human trials.

The first effective weight-loss drugs were amphetamines; by 1948, about two-thirds of weight-loss patients in the United States were being prescribed these drugs. However, they have other actions: during the Second World War, American servicemen consumed 200 million amphetamine pills to combat fatigue, and after the war amphetamines were widely used as recreational performance enhancers.

Amphetamines activate neuronal systems that use noradrenaline and dopamine as neurotransmitters. One of these is the A2 group of noradrenaline-containing neurons in the nucleus of the solitary tract in the brainstem; this site receives information from the gut via the vagus nerve, and some of the A2 neurons project to the hypothalamus to regulate food intake. Amphetamines and related drugs affect appetite by mimicking the effects of activating this pathway. However, by other actions in the hypothalamus, amphetamines affect libido, alertness, and blood pressure; their side effects include paranoia, hallucinations, and delusions, and they carry a risk of addiction.

In 1959, phentermine, a derivative of amphetamine that is much less addictive, was approved by the Food and Drug Administration (FDA) for treating obesity, but only for short-term treatment; once patients stop taking phentermine they quickly regain the lost weight.[2]

As well as an input from noradrenaline neurons, the hypothalamus receives an input from neurons that use serotonin as a neurotransmitter, and in the 1970s, fenfluramine, which stimulates this pathway, was approved by the FDA. In obese rodents and humans, fenfluramine reduced hunger, but produced only a small weight loss. However, whereas phentermine caused insomnia, agitation, constipation, and euphoria, fenfluramine caused drowsiness, sedation, diarrhea, and depression. At the University of Rochester, Michael Weintraub realized that, because phentermine and fenfluramine acted in different ways, they might be more potent when given together, and their opposite side effects might "cancel out." In 1992, he published a study looking at these drugs in combination.[3] Those taking the combined medication for 34 weeks lost, on average, 16% of their initial weight, while the placebo group lost just 5%. The doses that were effective—60 mg of fenfluramine and 15 mg of phentermine—were lower than the doses of these drugs when given individually.

In 1996, just four years later, 6.6 million prescriptions of the so-called fen-phen cocktail were issued in the United States. In the same year, the FDA approved dexfenfluramine, a drug closely related to fenfluramine, and this too was used mainly in combination with phentermine. The safety of these drugs in combination had not been tested: once the FDA has approved a drug, doctors in the US are free to prescribe according to their clinical judgment. The use of the two drugs in combination was "off label."

Then, in 1996, a study in the *New England Journal of Medicine* warned that the incidence of pulmonary hypertension was 23 times higher in patients taking the drug combination.[4] This affected just one in every 17,000 patients, but it is serious; stopping the medication might not stop the disease, and a lung or a heart-lung transplant might be needed. The following year, researchers at the Mayo Clinic reported 24 cases of a rare heart valve disease in women taking fen-phen. The FDA had already received nine similar reports, and now the agency asked all health care professionals to report such cases. This produced another 66 reports, so in 1997 the FDA asked drug manufacturers to withdraw fenfluramine and dexfenfluramine. By 1997, about 60 million people had been prescribed these drugs, mostly in Europe,[5] and when fen-phen was withdrawn there was little to turn to, and a scramble ensued to find alternatives.

Fenfluramine stimulates serotonin release from neurons. When serotonin is released, its duration of action is determined by how quickly it can be recycled: transporter molecules return it to the nerve endings to be reused, and sibutramine inhibits this mechanism. Sibutramine was approved by the FDA in 1997, but by 2010 it too had been withdrawn in most countries because of safety concerns. Lorcaserin, an antagonist of the serotonin receptors expressed by appetite-regulating neurons in the arcuate nucleus, is still currently approved for treating obesity, but it has only modest effects.[6]

Another potential treatment emerged from studies of how cannabis acts on the brain. When cannabis became a part of hippie culture in the 1960s and '70s, one side effect was "the munchies"—it stimulated appetite. Cannabis mimics the actions of naturally produced endocannabinoids, which are made in many neurons in the brain and which act at specific cannabinoid receptors. It became possible to make antagonists that block these receptors, and in the early twenty-first century, one such antagonist, *rimonabant*, was approved for use in treating obesity in Europe under the trade name Accomplia. However, in 2008 the European Medicines Agency recommended that doctors should no longer prescribe rimonabant because of the risk of serious psychiatric problems—including suicidal tendencies.[7]

Every current anti-obesity medication targets the brain except one. The exception is orlistat (marketed as a prescription drug under the name Xenical), which stops fats from being absorbed. In patients taking orlistat, about a third of the fat that would otherwise have been absorbed passes straight through the bowel; they need to go to the toilet more often and sometimes urgently, and they produce unpleasant fatty stools. They may also need vitamin supplements because vitamins that are soluble in fat are not properly absorbed when orlistat is taken. Unsurprisingly, there are problems with compliance: few patients tolerate the side effects for long.

To develop an effective drug with no adverse side effects is not easy. Anti-obesity drugs interfere with the normal mechanisms of appetite and metabolism, not with a disease process. The control of appetite is fundamental to our survival, and the mechanisms responsible are ancient, robust, and complex, interlinked not only with the regulation of body composition, metabolism, energy expenditure, and glucose homeostasis, but also with other fundamental drives, including the drive to reproduce. The reciprocal nature of the control of sexual behavior and appetite is conserved throughout evolution, and most drugs that affect appetite have effects on sexual behavior. It's hard to see how this can be avoided, because the same neurons in the hypothalamus control both.

Some years ago, a pharmaceutical company invited me as an adviser to a conference on peptide effects on feeding. At a moment when my attention had faded and my energy levels were low, I was asked about studies presented that morning. As far as I could see, the only observable outcomes from the experiments reported were that the animals might eat more, or less, or the same: perhaps they might have preferred to have sex, but that choice was not on the table. It was a flippant response, but more prescient than it deserved, as many of those agents did indeed turn out to affect sexual behavior.

A good example is the case of α-MSH, the most potent inhibitor of appetite so far known.[8] α-MSH is produced by the neurons of the arcuate nucleus that express the serotonin receptors that are targeted by drugs like fenfluramine, sibutramine, and lorcaserin. The α-MSH neurons spray axons to every nucleus in the hypothalamus and to most adjacent regions, and also to some distant sites in the brainstem and spinal cord. In the hypothalamus, they project to the paraventricular nucleus and supraoptic nucleus, where they contact oxytocin cells that regulate both appetite and reproductive behaviors. The axons do not go everywhere, but two types of receptor for α-MSH—MC3 and MC4 receptors—are expressed in every part of the brain.

In the 1990s, pharmaceutical companies were developing agonists for MC3 and MC4 receptors in an effort to develop new treatments for obesity. These efforts foundered in part because of an inability to influence appetite without simultaneously affecting sexual responses. When tested on male volunteers, MC4 agonists inhibited appetite but also promoted penile erection.

The names given to peptides are often disconcerting. The MSH in α-MSH stands for melanocyte-stimulating hormone. The functions of the α-MSH that is made in the brain have nothing to do with melanocytes, but α-MSH is also made in the intermediate lobe of the pituitary. As its name quite reasonably suggests, when α-MSH is secreted from the pituitary into the blood it stimulates melanocytes of the skin to produce melanin—it is the "tanning hormone." This effect is mediated by the MC1 receptor; the gene for this receptor is polymorphic, and some of the variants are associated with red hair, fair skin, and poor tanning ability, and carry a greater risk of melanoma.[9] For most of us, exposure to sunlight is a hazard that we manage with creams and common sense, but for patients with erythropoietic protoporphyria, even indoor lighting can be painful. This rare disorder arises from an enzyme deficiency that leads to the accumulation in blood of protoporphyrin, a photosensitive molecule. Melanotan I, a synthetic analog of α-MSH, is now available for these patients, for whom increased melanin offers an important protection.[10]

At high doses, melanotan II, another analog of α-MSH, has effects on the brain, as was discovered by Mac Hadley.[11] Hadley tested, on himself, the ability of melanotan II to produce a tan, but he used what proved to be an excessive dose, a dose that affected MC3 and MC4 receptors as well as MC1 receptors. He described the consequences in an account in which we can recognize both satiety-inducing effects and effects on libido:

> MTII caused a rather immediate, unexpected response: nausea and, to my great surprise, an erection (no figure provided). While I lay in bed with an emesis pan close by, I had an unrelenting erection (about 8 h duration) which could not be subdued even with a cold pack. When my wife came upon the scene, she proclaimed that I "must be crazy." In response, I raised my arm feebly into the air and answered, "I think we may become rich."

The behavioral effects of activating MC4 receptors—inhibition of appetite and stimulation of libido—are like those seen when oxytocin is delivered into the brain. This is no coincidence. The oxytocin cells express MC4 receptors and are targeted by axons from the α-MSH neurons, and activation of these receptors causes oxytocin release from dendrites. One site of action of this oxytocin is the ventromedial nucleus of the hypothalamus, which controls both appetite and sexual behavior: for any animal, finding food is an imperative when energy stores are low, but when this need is fulfilled the drive to reproduce becomes dominant. The α-MSH neurons are activated after eating, and they suppress appetite while stimulating sexual interest. Hunger impairs the willingness of a male rat to mate with a female, but if α-MSH is injected into his brain to mimic the effects of eating to satiation, his enthusiasm to mate is restored.[11]

The reciprocal regulation of appetite and sexual behavior and the roles of neuropeptides in both drives have ancient origins. The nematode *Caenorhabditis elegans* (*C. elegans*) is small and harmless; one of the simplest animals with a nervous system, and of no known economic importance. There are two sexes: a self-fertilizing hermaphrodite and a male. Of the 959 cells in a hermaphrodite, 302 are neurons, and the wiring diagram of these (the *connectome*) has been fully mapped; these neurons use many different signals, including about 100 different peptides. The worms move with the aid of 81 muscle cells, which generate undulations, and they find food by the smell of bacteria and from other clues to their presence, such as oxygen concentrations. *C. elegans* was the first multicellular organism to have its entire genome sequenced, and anyone who expected that the genome would be simple was in for a shock. The development and function of this organism is encoded by about 20,000 protein-coding genes—about the same number as in humans. At least a third of the genes have mammalian homologs, and many have remarkably conserved functions.

When crawling on a "lawn" of bacteria, *C. elegans* alternate between "roaming" to hunt for new food, and "dwelling" to exploit locally available food. Well-fed hermaphrodites rarely leave the local food patch, and this preference to stay depends on signaling at the *npr-1* receptor, which is homologous to the mammalian NPY receptor. In mammals, NPY neurons in the arcuate nucleus are essential for appetite regulation. When energy stores are low, the NPY neurons are activated and potently inhibit the α-MSH neurons, driving feeding behavior.

In *C. elegans,* the switch between roaming and dwelling is regulated by serotonin neurons, which promote dwelling, and neurons that produce a neuropeptide called PDF (pigment-dispersing factor), which promote roaming.[13] As already mentioned, serotonin also has important roles in appetite regulation in mammals.[14] The mammalian homolog of PDF is *vasoactive intestinal peptide,* which controls many aspects of gut function, but is also expressed in the suprachiasmatic nucleus of the hypothalamus, where it is implicated in the circadian regulation of appetite.[15]

Reproductive success in *C. elegans* is facilitated by mate-searching behavior, which is also governed by the PDF neurons. Males will leave an abundant source of food to explore their environment if no potential mates are present—but only when fully fed. This behavior is governed by internal signals that indicate the nutritional status and the reproductive state of the male, and these are mediated by an insulin-like signaling pathway and by signaling through a steroid hormone receptor. Thus, the same set of peptides and receptors is implicated in both feeding behavior and appetite in species as different as *Homo sapiens* and *C. elegans.*

Before the human genome project, it was assumed that our hundreds of thousands of genes would yield precise targets for developing new, selective drugs that would intervene in specific ways to correct imbalances in our physiology. But we don't have hundreds of thousands of genes; like *C. elegans,* we have about 20,000. Most of our genes do many different things in different cells, and, in the brain, most of our cells do many different things. Finding new drugs that are without serious side effects is hard.

In this vacuum, quack remedies from herbalists, homeopaths, naturopaths, and others abound, exploiting the desperation of patients. They don't work, and I'm not going to waste time on them. One thing does work: gastric bypass surgery. This surgery, which reduces the volume of the stomach, is almost magically effective both in producing weight loss in the most obese patients, and unexpectedly, in correcting type 2 diabetes.[16] We still don't understand why it is so effective. Originally, it was assumed that reducing the volume of the stomach would reduce how much food was absorbed, but now it is clear that this does not explain the outcomes. The surgery alters the physiology of weight regulation by changing the signals that go from the gut to the hypothalamus. These signals control our appetite and how we regulate plasma concentrations of glucose, and they even affect our choice of foods to eat.

Gastric bypass surgery is offered only to patients for whom other interventions have failed. It has serious risks: in 2004/5, a study reported that 1 in 200 patients died from complications within six months of surgery—fen-phen was withdrawn because of complications that affected just 1 in 17,000 patients, but the regulatory regime for new drugs is much more stringent than for surgical treatments. Gastric bypass is a treatment of last resort, but its effectiveness carries an important message—that the communication between the stomach and the brain can be modified to have a beneficial impact. That message drives much of the search for new drugs.

Several new hormones that are secreted from the gut have been discovered. Among these is glucagon-like peptide 1 (GLP-1), secreted from the small intestine, which has interesting effects on glucose homeostasis as well as on food intake. The GLP-1 receptor is present in the gastrointestinal tract on the sensory endings of neurons of the vagus nerve. These neurons signal to the nucleus of the solitary tract, where GLP-1 is also expressed in neurons that project to the paraventricular nucleus: thus, there seems to be a chain of GLP-1 signaling from the gut to the hypothalamus that is responsible for effects of GLP-1 on both body weight and glucose homeostasis.[17] Liraglutide, a long-acting GLP-1 analog, has been approved for the treatment of type 2 diabetes, and it produces a significant but still quite modest weight loss in adults.

Another anorectic gut hormone is peptide YY; this is secreted after a meal and it inhibits appetite by inhibiting the NPY neurons of the arcuate nucleus: these neurons have fibers that lie outside the blood-brain barrier, so peptide YY can reach them directly from the blood. Oxyntomodulin, secreted from the colon, has many effects, still incompletely understood. It reduces food intake, possibly by acting at GLP-1 receptors, but it also raises body temperature and so increases energy expenditure, and it also affects insulin secretion from the pancreas and gastric emptying.[17]

Because these hormones are involved specifically in communication between gut and brain, there is a reasonable hope that drugs based on them will be selective in their actions. Because peptide YY, oxyntomodulin, and GLP-1 all affect body weight but by different mechanisms, giving these in combination is likely to be more effective than any one alone, allowing lower doses of each to be used. This would reduce problems of receptor desensitization, and might also reduce the risk of adverse side effects. Steve Bloom and his colleagues at Imperial College, London, have long been studying gut hormones as potential treatments for obesity, and they have reported that the combination of peptide YY, GLP-1, and oxyntomodulin is indeed very effective for weight loss in humans.[18,19]

Other drugs being evaluated include agonists that act at MC4 receptors in novel ways. Peptides typically act at *G protein–coupled receptors*; when a peptide binds to one of these, the conformation of the receptor molecule changes, resulting in activation of a G protein that is associated with the receptor. Depending on the G protein and on other features specific to particular cells, this can result in the activation of one or more intracellular signaling pathways, with diverse effects, including on gene expression and on neuronal excitability. Because different agonists at a given receptor can produce different changes in receptor conformation, they can have different effects, and exploiting this variability is one potential way of minimizing side effects.

Another drug being used in combination therapy is naltrexone, an antagonist of opiate receptors—the receptors through which morphine and heroin act. The α-MSH neurons express these receptors, and as well as releasing α-MSH they also release β-endorphin that acts at opiate receptors as a negative-feedback signal, so blocking this receptor can enhance α-MSH release, amplifying satiety. However, opiate receptors are widely distributed in the brain, and there is natural concern about the possibility of side effects on mood.

Nobody envisages that indefinitely continued drug treatment will be a good way of treating obesity; the intention behind treatment is to restore a normal body weight but also to reset the hypothalamic appetite-regulating systems in a way that enables those systems to efficiently maintain this new body weight: how that might occur is the subject of the next chapter.

There are other things we might do about the obesity epidemic. We've banned smoking in restaurants and public places; we might do the same for food. Ban food advertising, ban anything that makes food tastier, close restaurants, clamp down on eating in public, sack celebrity chefs. McDonald's made food cheap, safe, and enticing; let's close them down. We won't do any of those things; eating isn't something we do in the search for happiness—it's one of those things that *is* happiness. Stopping people doing things that bring them happiness generally doesn't work, and if it does, perhaps it shouldn't.

18 Appetite

Obesity is a disease that different people are more or less prone to because of their genes. Conspicuous gluttony can cause obesity, but most obese people eat little differently from those of normal weight. Regular exercise will keep fat stores low and build muscle, but for an obese person, exercise poses a strain on an already stressed heart and joints, adds to hunger, and has little effect on body weight. Exercise is a good way of avoiding becoming overweight, and is great for the heart; it might also work as a punishment for obesity, but is probably not much use as a cure. It depends on your socioeconomic status; if you are poor, you will buy food that is cheap and rewarding, and those foods are likely to be energy dense. We're not helpless, but nobody chooses to be obese, and once we reach that state, it can be very, very hard to get back.[1]

What is apparent in the search for new treatments is that we still do not know enough about how we regulate appetite, metabolism, and body composition. Basic scientists came to these problems late: before 1994, there was little interest in how feeding was regulated. We got hungry, we ate, what was there to know? One of the few who recognized that there was something important to know was G. R. Hervey, who noted that an adult animal's intake and expenditure of energy are normally almost equal over long periods. In a 1969 paper in *Nature*[2] he quoted from a lecture by Reg Passmore, a nutritional physiologist then working at the University of Edinburgh. Passmore had pointed out that between the ages of 25 and 65 an average woman gains 11 kg. Over this time she eats roughly twenty tons of food; her weight change thus corresponds to an average daily error of only 350 mg of food, as compared with the exact amount required for energy balance.

I also have gained about 11 kg over the last forty years. Most of us do not weigh our food every day to see how many calories we're consuming. We eat heartily one day, the next day we may snack, or fall ill and starve ourselves. We binge intermittently, exercise intermittently, skip meals, drink (occasionally excessively)—we don't plan our diets. Yet, our body weights generally stay remarkably stable; we gain weight as we age, but usually very slowly. Gaining just one pound (450 g) a year will, over forty years, make any of us obese. One pound is roughly the equivalent of 3,500 calories; that's just 10 excess calories a day—less than a banana a week.

For Hervey, it was inconceivable that body weight could remain so constant by chance. He noted that adult rats are remarkably efficient at adjusting their food intake to provide a nearly constant supply of calories when the food was diluted with inert material, but he thought it unlikely that rats have a calorie counting mechanism. He thought that there must be some signal from the body's fat stores that regulates appetite by acting on the hypothalamus. He focused his attention on the ventromedial nucleus of the hypothalamus, as lesions to this small region were known to cause obesity in rats due, mainly, to an increase in the amount of food eaten. In an extraordinary experiment, he surgically joined pairs of rats so that their blood circulations were intermingled in a procedure known as parabiosis, and then lesioned the ventromedial nucleus in one rat of each pair. The lesioned rat as expected began to eat voraciously and gained weight rapidly; as it did so, the unlesioned partner ate less and grew a bit thinner, but the weight loss was quite slight.

Hervey's paper went almost unnoticed: in the next ten years it was cited just 16 times. His study was replicated by Han and colleagues, who again showed only slight weight loss in the unlesioned rats, a change that was not significant in their study.[3] Han et al. concluded that "humoral signals for inhibiting and facilitating feeding, if they do exist, cannot be transferred readily across the parabiotic union" and that Hervey's inference was "probably not correct." Han et al. were wrong and Hervey was right, but it would be another thirty years before this became apparent.

Our genes have an important influence on body weight; classic twin studies, using pairs of identical and fraternal twins, indicate that 40% to 70% of variation in body size is due to genetic factors.[4] Of course, what we inherit from our parents is just a predisposition to put on weight; whether we realize that predisposition is partly under our control, or perhaps more accurately under the influence of environmental factors. Knowing this, researchers had looked for animal models of genetically transmitted obesity by breeding from rats or mice that were fat. These programs produced two interesting strains: the *ob/ob* mouse, and the *db/db* mouse. These mice were very fat, and the *db/db* mouse was especially prone to severe type 2 diabetes (hence its name). Just a single gene was defective in the *ob/ob* mouse, and only mice homozygous for the affected gene became obese. Similarly, a single gene defect was involved in the *db/db* mouse, and again only homozygous mice became obese. But the defects were in different genes.

At the Jackson Laboratory in Maine, Douglas Coleman became interested in the fact that, while both strains were similarly obese, only the *db/db* mouse became severely diabetic. He speculated that some circulating factor might be affecting the secretion of insulin. If this factor were carried through the blood, he realized that he could test for its presence by linking the blood supply of the *db/db* mouse with that of a normal mouse using parabiosis. He expected that the normal mice might become diabetic, or that the diabetes of the *db/db* mouse might be ameliorated, but instead, the normal mice stopped eating. After about one week, their blood sugar concentrations had declined to starvation levels while their diabetic partners continued to overeat. He concluded that "the normal partner's failure to eat may be that, when a normal animal overeats, a substance is released which affects the satiety centers regulating food consumption. This substance may be released by, but be ineffective in, the diabetic mouse."[5]

Was the normal mouse starving because it was being dragged around by the larger *db/db* partner with little chance to eat? When Coleman paired *db/db* mice with *ob/ob* mice, which were of the same weight, the *ob/ob* mice responded in the same way as the normal mice had: they stopped eating. Coleman concluded that the *db/db* mouse overproduced a satiety factor but could not respond to it—perhaps because of a defective receptor—while the *ob/ob* mouse could respond to the factor but did not produce it.[6]

After Coleman's experiments, Hervey, with his graduate student Ruth Harris and others, went back to do one more parabiosis experiment, on genetically obese Zucker rats.[7] When Zucker rats were paired parabiotically with lean partners, the lean partners reduced their body fat content by almost 50% compared with members of lean-lean pairs; like the *db/db* mice, it appeared that Zucker rats produced a factor that inhibited appetite but could not themselves respond to it.[8]

Still, most of the scientific community remained unconvinced. In the twenty years that followed his experiments Coleman's papers were scarcely cited. As Coleman later reflected, "Despite these clear results, many of my colleagues and many in the obesity field maintained the dogma that obesity is entirely behavioral, not physiological."[9]

Some workers continued to pursue the possibility that there was a regulatory system in the body that homeostatically controlled body weight. Infusions of glucose, glycerol, and insulin all would reduce food intake, but none of these really satisfied the conditions for a homeostatic regulatory system. To postulate a homeostatic regulation of body weight implies that there is a predefined "set point" that the system works to maintain, and that the appetite-regulating centers of the brain are sensitive to a reporter of fat mass, but what this might be was unclear. The average concentration of insulin in plasma correlates with fat mass, but it fluctuates markedly over a day, and rather little enters the brain. Many factors can increase or decrease hunger or satiety, but, as Kissileff noted in 1991, apparent regulations "may be the result of simple averaging processes. They may be no more complicated than the rule that dividing a large group of random numbers by the size of the group will give very reproducible averages."[10]

When the gene affected in the *ob/ob* mouse was identified in 1994, the story made the cover of *Nature.*[11] The headline was "Mouse Weighed Down by Genetics," and the cover showed an *ob/ob* mouse on a set of weighing scales counterbalanced by two normal mice. The paper reported the discovery of the first new hormone in fifty years, and the authors named it "leptin" from the Greek *leptos,* meaning "thin." Leptin is produced by adipocytes and it signals to the brain, especially to the α-MSH neurons in the arcuate nucleus. Leptin is secreted in proportion to body fat mass; it is the body's way of telling the brain how much energy is in its store, and it controls our appetite in a very appropriate way—when our energy stores are low, we get more hungry, when they are high, less so. The *ob/ob* mouse produces no leptin: however fat it gets, its hypothalamus fails to recognize this, and acts as though the mouse is starving. Give an *ob/ob* mouse leptin and its body weight soon falls to normal.

When the *db/db* mouse defect was identified, the picture seemed complete. This mouse produces leptin in abundance but has no leptin receptors. However much leptin is produced the hypothalamus can't see it, so like the *ob/ob* mouse the *db/db* mouse eats as though it's starving—which is exactly what its hypothalamus is telling it. Then, the defect in the obese Zucker rat was identified: like the *db/db* mouse, the mutation was in the leptin receptor.[12] The discovery of leptin and of its actions in the brain had confirmed Coleman's prediction, made twenty-five years earlier.

That we don't have to think about planning our meals tells us that our brains do the calculations for us unconsciously and compensate when we eat more or less than we need. Surely all we needed was a drug that mimicked leptin, and we'd shrink our appetite. Sadly, it didn't work out. A few people are obese because they, like the *ob/ob* mouse, don't produce leptin, but in humans this defect is extremely rare. Those affected, like the *ob/ob* mouse, eat voraciously and are very obese, and for these patients leptin is indeed a wonder cure. But most obese people have *high* levels of leptin in their circulation. This shouldn't be surprising: leptin is produced by adipose tissue, so the fatter you are, the more leptin you produce. This means that people are usually obese *despite* high levels of leptin, not because they have low levels.

This is key to why we think of obesity as a disease. Obese individuals are generally insensitive to leptin, just as type 2 diabetics are to insulin. This isn't a glib analogy, because the biochemical mechanisms involved in leptin resistance are like those involved in insulin resistance. Just as treating insulin resistance is difficult, so is treating leptin resistance. If we class type 2 diabetes as a disease, we can't consistently deny that obesity is a disease.

In 1999, another new hormone, ghrelin, was discovered.[13] Ghrelin is released from the empty stomach: its levels fall after a meal and rise between meals. It too acts on the hypothalamus, on the NPY/AgRP neurons that drive food intake, and it is the most potent hunger-inducing signal so far known.

Now we can give an explanation of how obesity develops. It's a narrative with gaps and flaws—a narrative, based on our current understanding, that will change as we learn more and that may be wrong in important ways—but this is where we are now.

In the natural, animal world, obesity is rare. When food is abundant, animals reproduce rapidly, leading to a balance between available food and population size. When animals compete for food, the rules are simple: when you're hungry, look for food and eat when you can; when you're sated, find a mate and breed.

We, in the developed world, rarely experience real hunger. We eat because it's dinnertime, because we're stressed, because we like the taste, because it's the sociable thing to do, because we're bored, because it's Christmas. It's surprising that we ever stop eating, but we do, because although we rarely experience hunger we still experience satiety—the feeling of fullness and satisfaction after eating—and because our appetite is still moderated by signals, like leptin, that restrain us.

There are two systems that control appetite: the homeostatic system that strives to regulate energy stores; and a hedonic system. The homeostatic system is sensitive to leptin and ghrelin, and also to insulin and many other hormonal signals and neural signals from the gut. Ghrelin from the empty stomach reaches high levels after a fast, and it activates neurons in the arcuate nucleus that make NPY. Leptin, secreted by adipocytes, reports on the body's fat reserves; it inhibits NPY neurons, while activating other neurons that express anorexigenic factors, notably neurons that express α-MSH. These are reciprocally linked with the NPY neurons, and which population is dominant determines how much an animal will eat. As an animal eats, neural and endocrine signals from the gut report on the volume ingested and on its composition, including its complement of fat, carbohydrates, and protein. These signals, relayed by "satiety" centers of the caudal brainstem, converge on the ghrelin- and leptin-sensing circuits of the hypothalamus. These in turn project to other limbic sites, including the paraventricular nucleus, which is the primary regulator of the sympathetic nervous system and which also regulates the hypothalamo-pituitary-adrenal axis—the "stress axis."

These pathways are powerful moderators of energy intake, but it's not just about appetite: when we eat more, we also spend more energy: we become more active, and our metabolic rate increases, and when we eat less we become lethargic and reduce our metabolic rate. Conversely, if we exercise more, our appetite increases, to provide more fuel for our needs. These uncontroversial facts tell us that you can become obese without eating noticeably differently from your lean neighbor; that restricting food intake can have disappointing effects on body weight; and that taking more exercise will make you fitter and healthier but might do little for your weight.

The hedonic system makes eating rewarding; it's why we prefer certain foods, why we value and enjoy taste—and maybe what makes food, for an unlucky few, addictive. The brain circuits involved in food reward overlap with those involved in other types of reward, such as that for of drugs of abuse, and those of gambling and alcohol that did for Bowed Joseph.

But the hedonic system isn't independent of the homeostatic system. Ghrelin doesn't simply trigger eating, it also enhances the reward value of foods, and can guide dietary choice, sometimes in unexpected ways. When rats are offered a choice of lard (100% fat), sucrose, and chow they increase their lard consumption. If they are then given ghrelin into the brain, they change their food choice and consume more chow. Interestingly, these effects are different from what is seen after fasting, when it is the consumption of energy-dense foods that is prioritized.[14] Several other gut- and fat-derived hormones also impact on food reward circuitry: leptin, for instance, affects how the dopamine neurons of the ventral tegmental area encode food reward.[15]

The homeostatic systems counterbalance the rewarding effects of eating. But the most rewarding foods are often dense in energy, often in a form that is easily absorbed, and are often cheap and readily available. The modern world is stressful, full of small but inescapable pressures; and we compensate by comfort eating—again, this isn't a glib analogy, because the hypothalamic centers that mediate our responses to stress are also sensitive to energy supply.[16] In a direct way, in a way that we can demonstrate in rats and mice just as in people, eating reduces our stress.

Even a small imbalance between energy intake and energy expenditure might, over time, cause the fat to build up—if we carry on eating more despite the restraining influence of leptin, and don't increase our energy expenditure, and keep the room thermostat turned up. Over time, the leptin signal starts to lose its force, and we're on the path to obesity. In the leptin-resistant state, we're in trouble. We might be fat, but our hypothalamus doesn't think so, and if we try to lose weight by dieting, it responds as though we were starving. The hunger signals—ghrelin and others—rage unabated by a leptin signal that is unheard.

When we're truly starving, our hypothalamus defends our body weight. Our metabolic rate slows, our body temperature falls, we become lethargic: we switch off body systems that don't matter so much in the short term. We turn down our reproductive systems: there's no point in reproducing when there's not enough food. We turn down the immune system: there's an acute challenge, starvation, now's not the time to invest in a long-term defense against the hypothetical threat of infection. We turn the hunger signals up: look for food, look for food, forget everything else.

The obese, leptin-resistant person on a low-calorie diet feels depressed, lethargic, cold; they lose libido, are prone to minor infections, and all the time have a gnawing hunger, and the pounds are slow to slide off. Leptin resistance is not easily reversed. After the end of a diet, weight that is lost is usually regained, and often exceeded. Diets generally don't work in the long term, and may even make things worse. Indeed, one of the strongest predictors of weight gain is weight-loss dieting. One of the biggest studies to demonstrate this was the Growing Up Today Study, a prospective study of more than 16,000 adolescents. At the three-year follow-up, adolescents who were frequent or infrequent dieters had gained significantly more weight than nondieters.[17] The longest study that demonstrates this is Project EAT (Eating and Activity in Teens and Young Adults), a population-based study of middle and high school students.[18] This study, which controlled for socioeconomic status and initial BMI, again showed that the strongest predictors of weight gain were dieting and unhealthy weight-control behaviors. The behaviors associated with the largest increases in BMI were skipping meals, eating very little, using food substitutes, and taking "diet pills."

An answer seems far away, and probably won't be a single answer but many partial answers. For now, surgery has a place, and when we understand a little better why surgery works, there may be new treatments that mimic the beneficial effects without the risks.

Our understanding of the role of food components is still poor. Certain foods, including fat and protein, potently activate the satiety pathways that limit meal size, while others, like carbohydrates, might disproportionately drive weight gain, but it's more complex. I've talked about ghrelin and leptin as though these are the most important hormones, but they were just the first to be discovered.[19,20] Adipocytes secrete about 100 different proteins including adiponectin, apelin, interleukin-1β, interleukin-6, interleukin-10, macrophage chemotactic protein 1, resistin, tumor necrosis factor-alpha, and transforming growth factor β. These include "autocrine factors," acting as feedbacks to the adipocytes, and "paracrine factors," acting on neighboring cells. Some are hormones; they act on the lung, skeletal muscle, heart, liver, and blood vessels and affect insulin sensitivity and secretion, fat distribution, lipid and glucose metabolism, endothelial function, blood pressure, hemostasis (control of bleeding), and immunity as well as appetite and energy expenditure.[21] The pancreas secretes amylin, glucagon, and pancreatic polypeptide as well as insulin; cholecystokinin and secretin are secreted from the duodenum, peptide YY from the distal small intestine, glucagon-like peptide from the large intestine, adrenomedullin and PAMP (proadrenomedullin N-terminal 20 peptide) from throughout the gastrointestinal tract. Some of these act on the hypothalamus, some, like adrenomedullin, act at the circumventricular organs, and others, like cholecystokinin, activate vagal inputs to the brainstem. There are also signals from muscle[22] and bone[23] to the energy-regulating systems of the brain, and what information they carry and how they affect energy balance we don't yet know. Nor does the information flow only from the periphery to the brain—adipocytes are themselves regulated by NPY released from sympathetic nerve terminals that innervate fat depots in the body, and by α-MSH secreted from the pituitary.[24] Our physiological understanding has not kept up with the pace of discovery, and there are gaps in our understanding of even the best-studied hormones—we have little idea, for example, of the mechanisms by which leptin and ghrelin are secreted.

Adding to the enormous complexity in the signals that reach the brain, there is further complexity in the signals within the brain that process this information. We class neuropeptides as *orexigenic* or *anorexigenic* depending on whether they increase or decrease feeding: NPY, AgRP, orexins A and B, galanin-like peptide-1 (GALP-1), melanin-concentrating hormone, dynorphin, and relaxin 3 are among the former; α-MSH, prolactin-releasing peptide, CART, cholecystokinin, nesfatin, corticotropin-releasing factor, and oxytocin are among the latter.

If we ask ourselves what makes us eat, we might answer in many different ways. We eat because our stores of energy are low, or perhaps because we expect to be hungry or to need energy in the near future. Or because what is available is particularly attractive, because we are stressed and looking for distraction, or because we have an uncomfortable feeling from an empty stomach—or perhaps we need salt or some other nutrient, or it is dinnertime; and we might stop eating for at least as many reasons. So what do we learn from knowing that a peptide has an effect on food intake?

There are two right answers: nothing much; and something important. From the simple fact of an effect on feeding we learn nothing much about the physiological role of the peptide: there are so many ways that an effect in the brain might alter our motivation to eat. An effect on feeding is just a starting point: to study the role of any peptide on energy balance, we would now ask a few other questions: what effect does it have on energy expenditure—metabolic rate, thermoregulation, and locomotor activity? Does it affect meal size or meal frequency, or the circadian timing of feeding? Does it affect nutrient choice—fat versus carbohydrates versus protein? Does it affect fluid intake? Is the effect different in males and females? Is there an effect on taste, or on food reward? Until we properly understand the physiology, the advice we give about nutrition will continue to be rich in hubris but rest on flimsy scientific foundations.

19 The Sweet Hormone

Yes, oxytocin is in the list of peptides in the brain that affect appetite.[1] I'm not going to argue that it is the most important. I don't know how to distribute importance among the many neuronal contributors to energy balance and don't imagine that it would be useful or illuminating. I'm giving this chapter to oxytocin, mainly by way of introducing how it is that the same neurons can do several very different things.

In rats, central injections of oxytocin not only affect social and sexual behavior, they also potently inhibit appetite, as do central injections of mesotocin in chicks. As we saw in the previous chapter, the simple fact of an effect on feeding is only a starting point. The questions we ask when we investigate any peptide's role in energy balance must address *how* it changes our motivation to eat: whether it affects energy expenditure, pattern of intake, choice of nutrients, intake of fluid; whether it differentially affects females and males; whether it affects taste or the experience of food reward. In the case of oxytocin, the short answer is probably yes, to most if not all of the above.

Some oxytocin neurons in the paraventricular nucleus project to the spinal cord, with effects on thermogenesis, pain processing, and erectile function. Others innervate the dorsal vagal complex in the brainstem, where there is a high density of oxytocin receptors: this region regulates the digestive functions of the stomach and esophagus, and it projects project back to the hypothalamus to stop eating when the stomach is full. Others project to the nucleus accumbens, part of the brain's "reward circuitry": there, oxytocin suppresses reward-driven food intake while enhancing the reward associated with social interactions. Transgenic mice that lack oxytocin or its receptor eat about the same amount as wild-type mice, but they show a greater preference for sweet-tasting solutions than wild-type mice, and they are more likely to become overweight.

These different actions of oxytocin are not all hard to explain; some are mediated by different subpopulations of oxytocin neurons that might have properties specialized for those different roles, and which project to different brain sites where the oxytocin they release can act relatively locally. Much more interesting is how the magnocellular oxytocin neurons, all of which are involved in milk ejection and parturition, can also have a role in appetite.

For me, this story came to life when Louise Johnstone in my lab set about using c-*fos* protein as a marker to determine which neurons in the brain are activated by feeding: c-*fos* is an "immediate early gene" that neurons begin to express when their activity increases above "normal" levels.[2] We had expected that in a hungry rat, the neurons that drive feeding would be active and express c-*fos*, while neurons that mediate satiety would be active only at the end of a meal. Johnstone allowed rats food for just two hours each day, always at the same time. The rats soon learned to expect the arrival of food, and when it arrived they would eat voraciously for about 90 minutes before grooming themselves and going to sleep. Johnstone looked at the brains of rats killed before or at the expected time of food arrival, and at intervals after food had arrived. We had expected that the NPY neurons would already be active at the expected time of food arrival, stimulated by ghrelin from the empty stomach, and that the lateral hypothalamus, a "hunger center" that contains other appetite-stimulating peptides like orexin and MCH, would be active too. Equally, we had expected that the α-MSH neurons and brain areas involved in satiety would become active toward the end of the meal.

This is not what Johnstone saw. She saw little activation of any neurons before or at the expected time of food arrival, but strong activation in many areas after 30 minutes of feeding. Because there is a delay of about 30 minutes before any stimulus that stimulates c-*fos* expression can result in the appearance of c-*fos* protein, these neurons must have been activated as soon as food appeared. The levels of c-*fos* protein continued to rise, peaking in most areas at about 90 minutes before beginning to slowly decline after the rats stopped eating. It seemed that the neurons driving feeding were activated by the arrival of food, perhaps by its smell and taste. This was perhaps not so surprising: after all, we begin banquets with appetizers to pique our appetite.

But there was no difference in the timing of appearance of c-*fos* protein between neurons involved in stimulating feeding and those involved in satiety. The populations were activated a simultaneously. The satiety pathways were also engaged as soon as the rats started eating. Perhaps this too should not have surprised us—the signals from the gut that inhibit feeding begin to be released as soon as food arrives there.

Was it really the arrival of food that triggered this response in the brain? Johnstone took other rats and killed them an hour after the time of expected arrival of food but without giving them any food. These showed the same activation of c-*fos* in areas of the hypothalamus involved in hunger—the neurons were activated not by the taste or smell of food but by the expectation of food arriving. The lateral hypothalamus was even more strongly activated in rats that were not fed. Just as we learn to become hungry as dinnertime approaches, so these rats were intensely hungry when an expected meal did not appear. However, none of the brain regions involved in satiety were activated: the nucleus of the solitary tract was quiet, as were the α-MSH cells.

Of all the brain areas that Johnstone looked at, the strongest activation was in the supraoptic nucleus, where feeding evoked strong expression of c-*fos*. This activation, like that in the nucleus of the solitary tract and in α-MSH neurons was present only in fed rats, never in rats that were expecting but did not receive food.

The results were, for us, the strongest indication yet that magnocellular oxytocin neurons are involved in appetite. There was, however, already circumstantial evidence that magnocellular neurons might be involved in appetite. These neurons receive an innervation from the nucleus of the solitary tract, and in rats, gastric distension and intravenous injections of cholecystokinin strongly activate them. Cholecystokinin is a hormone secreted from the duodenum in response to food ingestion. It acts on the sensory endings of afferent neurons of the gastric vagus nerve, and this effect is relayed by the nucleus of the solitary tract, by neurons that produce noradrenaline and others that expresses prolactin-releasing peptide (as is often the case with the names of neuropeptides, this name is inappropriate). However, both cholecystokinin and gastric distension were thought to be stressful, and as it was known that other types of stressors would activate magnocellular oxytocin neurons, these findings did not unequivocally suggest a specific role in appetite regulation.

Then, two years after Johnstone published her paper, Giles Yeo and his colleagues in Cambridge showed that leptin strongly affects the oxytocin system.[3] They were searching for genes involved in mediating the effects of leptin, and reasoned that such genes would be downregulated in fasted mice when leptin levels were low, and upregulated again when the fasted mice were treated with leptin. They used laser-capture microdissection to cut out just the paraventricular nucleus from brain sections, isolated the RNA from these samples, amplified the RNA, and used a microarray to measure gene expression. In all, 89 genes met stringent criteria to be considered as leptin-regulated genes. The 25 genes most strongly influenced included three secreted peptides: thyrotropin-releasing hormone, oxytocin, and vasopressin. Although the paraventricular nucleus contains both parvocellular and magnocellular oxytocin neurons, the magnocellular neurons are more abundant and express much more oxytocin. To account for the Cambridge results, it seemed that they must be affected by fasting and leptin.

We went back to the supraoptic nucleus to study oxytocin neurons, and found that fasting reduced their spike activity while systemically administered leptin enhanced it. By then, more evidence had accumulated. The oxytocin neurons were responsive to glucose and insulin. They expressed other appetite-inhibiting factors, including cholecystokinin itself and nesfatin. They were activated by oleoylethanolamide (a hypophagic lipid-amide released by the small intestine in response to fat intake), by oral delivery of sweet solutions, and by delivery of food into the stomach by gavage. It was clear that many different appetite-related signals converge on these neurons.

The α-MSH neurons innervate both the paraventricular nucleus and the supraoptic nucleus, which densely express the MC3 and MC4 receptors through which α-MSH acts. In oxytocin neurons, α-MSH induces mobilization of intracellular calcium stores, expression of c-*fos*, and dendritic oxytocin release, but it also inhibits spiking activity and therefore inhibits secretion into the blood. This inhibition is the consequence of the evoked production of endocannabinoids, which suppress the release of neurotransmitters from afferent endings that express cannabinoid receptors. Interestingly, in pregnancy α-MSH has no effect on oxytocin neurons, and while central injections of α-MSH increase c-*fos* expression in the paraventricular and supraoptic in nonpregnant rats, this effect of α-MSH is also suppressed in pregnant rats. Perhaps this contributes to the hyperphagia of pregnancy: while mammals generally maintain a very stable body weight, in the last third of pregnancy they gain fat mass through increasing their food intake.

The magnocellular oxytocin neurons are the source of oxytocin in the plasma, but they also release abundant oxytocin from their dendrites, and this acts at two relatively close sites that express abundant oxytocin receptors: the amygdala, which contains only a few oxytocin-containing fibers, and the ventromedial nucleus, which appears to contain none. The ventromedial nucleus is important for glucose homeostasis, and it also controls sexual behavior, feeding, fear behavior, and aggression. These behaviors are not mutually compatible: given the motivation and opportunity to have sex and to eat, animals generally do one or the other, unless they are afraid, in which case they may fight or flee but are unlikely to eat or mate. Oxytocin enhances sexual behavior while suppressing feeding and fear, so, by its actions at the ventromedial nucleus, it is important for fundamental behavioral decisions.

In rodents, oxytocin secretion induced by food intake promotes sodium excretion and influences gastric motility, but these effects are not seen in all mammals. In humans, oxytocin secretion is stimulated by exercise (and this is associated with altered fluid balance), but not by osmotic stimuli as it is in rodents, nor does oxytocin induce natriuresis. Oxytocin secretion in humans is not stimulated by gastric distension or by systemic administration of cholecystokinin; instead, these activate *vasopressin* secretion. The same is true in ferrets.[4] It seems that there may be a difference between species, like ferrets and humans, that can vomit when they overeat and those, like rodents, that can't.

Nevertheless, oxytocin and its receptor are present throughout the gastrointestinal tract in humans and other mammals. Oxytocin acts as a brake on intestinal motility, promotes the development and survival of enteric neurons and intestinal crypt cells, regulates the permeability of the mucosal lining of the intestine, and protects against inflammation. Oxytocin receptors are also present in the rat pancreas, where oxytocin can stimulate insulin and glucagon secretion. Adipocytes also have oxytocin receptors, and through these, oxytocin induces lipolysis—the breakdown of lipids into glycerol and free fatty acids. Oxytocin also increases the formation of osteoclasts, cells that break down bone and are essential for repair and remodeling of bone, and in mature osteoclasts it inhibits this bone resorption. Mice lacking oxytocin or its receptor develop osteoporosis, a condition in which bones become fragile, and this worsens with age. In rabbits, glucocorticoid-induced osteoporosis can be prevented by giving oxytocin systemically, and in mice, oxytocin can reverse ovariectomy-induced osteopenia and adiposity.

So now, to come to the point of this chapter—how can one hormone, oxytocin, be involved simultaneously in so many different things?

It's not hard to conceive that different subpopulations of oxytocin neurons do different things, if they project to very different places that are far apart. Nor is it hard to conceive how oxytocin produced at peripheral sites has specific, very local actions on adjacent cells—those cells might be indifferent to the oxytocin concentrations in the plasma but might receive enough oxytocin from adjacent cells to be activated. Harder to understand is how the magnocellular oxytocin neurons can do several apparently very different things simultaneously without conflict. To spell it out, these neurons, in the rat, regulate milk ejection, parturition, food intake, social and sexual behavior, and sodium excretion.

Even milk ejection isn't simple. In marsupials, milk ejection is regulated by mesotocin, the marsupial homolog of oxytocin; they rear their young in a pouch—a pouch that can contain two young of very different ages. Dennis Lincoln and Marilyn Renfrew studied milk ejection in the wallaby *Macropus agilis*[5] Marsupials give birth to young that are almost embryonic in form; for the first 100 days or so of life, they remain continuously attached to a nipple. By the time it is 200 days old, the infant will have left the pouch, returning intermittently to suckle, by which time a second newborn infant may be attached to a different nipple. So how can mesotocin secreted from the pituitary provide an appropriate supply of milk to each of these two young, whose patterns of suckling and whose needs are so very different? Lincoln and Renfrew showed that the mammary gland that is suckled by the newborn is exquisitely sensitive to mesotocin; this suckling leads to intermittent pulses of mesotocin, which, though small, will trigger milk ejection at this gland, but not at the other gland, suckled by the older sibling, which is less sensitive to mesotocin. The suckling of that older sibling is a stronger stimulus, and leads to occasional large pulses of mesotocin—these will cause milk letdown at both glands—but the gland suckled by the newborn responds similarly to this large pulse as to a small pulse. They therefore concluded that "the wallaby, far from discarding during its evolution the milk-ejection response to oxytocin, has refined the response to permit it to feed simultaneously both newly born pouch young (<1 g) and young at foot (>2,500 g)."[5]

The response of a tissue to oxytocin (or in this case mesotocin) is defined by the level of expression of oxytocin receptors in that tissue. Turning to how oxytocin can regulate both uterine contractions and milk ejection, the explanation is the same—the uterus is only fully sensitive to oxytocin at the time of parturition, and the mammary gland only during lactation.

But what about natriuresis? Sodium excretion must be regulated throughout life, including during lactation. Here it is the pattern of secretion that is critical. The rodent kidney is extremely sensitive to oxytocin, and at the kidney, natriuresis is regulated by a graded, continual release of oxytocin governed by the continual background activity of oxytocin neurons.[6] Natriuresis is a slow and continuing process of sodium excretion: pulses of oxytocin are neither here nor there, their impact on the kidney is too brief and intermittent to count. The mammary glands, by contrast, are relatively insensitive to oxytocin—the graded background release that regulates natriuresis is invisible to them, all that matters is those large pulses. So by regulating their background activity and generating intermittent pulses, magnocellular oxytocin neurons can fulfill both functions without conflict.

These neurons also regulate both feeding behavior and sexual behavior. There is no conflict between these two—oxytocin inhibits one while enhancing the other, but how is a conflict avoided between these and the other functions of oxytocin? Here we must recall that because dendritic secretion of oxytocin does not require spike activity, it can be regulated independently of oxytocin secretion from the posterior pituitary.

The magnocellular oxytocin neurons are supreme multitaskers, with multiple sensory properties and multiple functional roles—characteristics, as we will see, of the neurons of ancient organisms from which they evolved.

20 The Hybrid Neuron

In the multiplicity and diversity of the physiological effects produced by these various chemical messengers one is apt to lose sight of the fact that we are here investigating one of the fundamental means for the integration of the functions of the body. These are not merely interesting facts which form a pretty story, but they are pregnant of possibilities for our control of the processes of the body....

—E. H. Starling, The Harveian Oration, delivered before the Royal College of Physicians of London on St. Luke's Day, 1923[1]

The classical neuron has dendrites, where information is received from other neurons and converted into electrical signals; a cell body, where these signals are converted into spikes; and axons, which carry the spikes to synaptic endings where they trigger the release of neurotransmitters. Neurons differ in their morphology: some have exuberantly bushy dendrites; some have several axons that have many branches; magnocellular vasopressin and oxytocin cells have two or three long but unbranching dendrites and just one long axon. Typically, neurons transmit information at several thousand synapses, and receive several thousand inputs at their cell bodies and dendrites, but there is great heterogeneity—in the rat, about 12,000 dopamine neurons of the substantia nigra project to 2.8 million spiny neurons in the striatum with axons that are on average 40 cm long and which each give rise to between 100,000 and 250,000 synapses.

Membrane properties vary from cell to cell and between the compartments of a cell, and those properties influence how incoming information is processed and how outgoing information is patterned. Some dendrites have extensive signal-processing abilities and can generate spikes; others are passive conductors of synaptic inputs. Classically, synapses are between axon terminals and dendrites, but some connect axon terminals to axon terminals, or dendrites with dendrites. Nevertheless, the classical image remains dominant, conveying the key understanding that the main point of information exchange is the synapse.

The classical synapse has several distinctive features. The presynaptic ending contains about 5,000 small vesicles, each of which contains a few thousand molecules of neurotransmitter, usually glutamate or GABA, sometimes noradrenaline, serotonin, or dopamine, sometimes acetylcholine or histamine. There is continuous remodeling of the synaptic architecture; some synapses proliferate as others are eliminated, guided in part by activity in the presynaptic cell and in part by signals from the postsynaptic cell. This web of interneuronal communication is generally understood to be the basis of information processing in the brain.

But Kyell Fuxe has long argued that some of these "neurotransmitters"—notably serotonin, noradrenaline, and dopamine, act mainly at *extrasynaptic* receptors. Many of the neurons that release these neurotransmitters have very diffuse projections in the brain and spinal cord, and he proposed that they use a mode of signaling that he called "volume transmission." Noradrenaline, serotonin, and dopamine are all messengers that act close to their site of release if not purely at synapses, but because they are released simultaneously from diffusely distributed projections, they go indiscriminately to all neurons in a particular brain region. Fuxe suggested that such pathways deliver "emotional" signals to the brain—such as the "reward" signals that act to reinforce experience-dependent plasticity in neuronal circuits.[2]

The dopamine neurons of the substantia nigra have been studied intensively because it is their death that causes Parkinson's disease. The substantia nigra is a large structure in the midbrain that is involved in the control of movement and in reward: it gets its name because the abundant dopamine neurons there express a pigment, neuromelanin, giving it a dark color. The substantia nigra is reciprocally connected with the striatum, a large region in the forebrain that also receives major inputs from glutamate neurons of the cortex and thalamus, and its output cells are "medium-sized spiny neurons" that release GABA (spiny neurons are neurons whose dendrites exhibit a profusion of protruding "spines" that increase the surface area of the dendrite, providing more sites at which synapses can form). It was once thought that the dopamine projections from the substantia nigra to the striatum allow an exquisite cell-specific tuning of the responses of the spiny neurons to excitatory inputs from the cortex. The numbers tell another story. In the rat, the axon of each dopamine neuron innervates about 6% of the total area of the striatum.[3] These synapses were once thought to target the bases of dendritic spines, where they could specifically modulate the glutamate inputs from cortical projections. But Jonathan Moss and Paul Bolam in Oxford, in a detailed electron microscopic study, carefully quantified the contacts between dopamine terminals and their relationship to the terminals from cortical and thalamic projections. They found that "the organization of the dopaminergic nigrostriatal system is such that striatal neuropil is located within a dense, evenly spaced lattice-work of dopaminergic axons, and the probability of being apposed by a dopaminergic axon is principally dependent on the size of the structure. Thus, all similarly sized striatal structures have an approximately equal probability of being apposed by a dopaminergic axon. The organization is such that apposition by, and proximity to, a dopaminergic axon is random in nature."[4]

Thus while dopamine was once thought to have actions confined to the synapse, many dopamine receptors are found outside synapses, and dopamine spills over from synapses to act at these extrasynaptic receptors. Moss and Bolam found that every structure in the striatum was within 1 μm of a dopamine synapse. In short, dopamine in the substantia nigra looks like a diffuse and indiscriminate signal, not one of exquisite temporal and spatial selectivity—more like a hormone than a neurotransmitter.

Neurons are not the only cells in our bodies that process and transmit information. The classic endocrine systems include five major steroid hormones: testosterone, estrogen and progesterone secreted by the gonads, aldosterone and cortisol by the adrenal cortex. These hormones are not stored, but are produced on demand; because they are soluble in lipids, no cell membrane can contain them, and they regulate gene expression in tissues throughout the body, including in the brain—there is no blood-brain barrier to steroids. Prostaglandins are other lipid-soluble hormones that are not stored but are produced on demand. They come in diverse forms, are produced by most tissues in the body, and have diverse actions; they regulate blood flow in many tissues, and regulate uterine contractions in labor. Thyroid hormones are also lipid soluble, but are carried in the blood to their sites of action by binding proteins, and they regulate protein, fat, and carbohydrate metabolism and brain development.

Steroid hormones are soluble in lipids and so generally have no difficulty in accessing the brain, but there are some important exceptions. One of these exceptions is important for understanding the difference between the brain of a male and that of a female.

The mammalian hypothalamus is sexually dimorphic; the hypothalamus differs between males and females in a number of very visible ways. In the human hypothalamus, the *sexually dimorphic nucleus* is a part of the preoptic area that is twice as large in men as in women, and many other differences are apparent on close examination. In part, the sexual dimorphism arises from the effects of sex steroids released in early postnatal life, effects that last throughout life.

A newborn male mammal has large testes; for a short while after birth, the testes secrete large amounts of testosterone, and its actions "masculinize" the developing hypothalamus: if the testes of a rat are removed soon after birth, the hypothalamus develops as a "female" hypothalamus. If an ovary is transplanted into an adult male rat that was castrated at birth, the rat will show cyclic ovulation, and at the estrous phase of that cycle will respond to a male with a *lordosis reflex*—the arched-back posture that invites mounting by the male. The same is not true of a male rat castrated in later life: the ovulatory mechanism is potentially present in the male brain at birth, but is prevented from developing by the actions of testosterone at a critical period in neonatal development. Conversely, if a newborn female rat is given an injection of testosterone in her first few days of life, the ovulatory mechanism of the hypothalamus fails to develop and she will be forever sterile.

What does testosterone do that makes such a profound difference to the developing hypothalamus? Strangely, it is not testosterone that masculinizes the male brain, but the female sex hormone, estrogen. If a large amount of estrogen is injected directly into the brain of a newborn female rat, she will develop a male hypothalamus just as if she had been injected systemically with testosterone. The female is normally protected from the relatively low amounts of estrogen that she experiences in neonatal life because the fetus makes a plasma protein called alpha-fetoprotein, which avidly binds estrogen and prevents it from reaching the brain.

In males, by contrast, testosterone is abundant and is less effectively bound in plasma. When testosterone reaches the hypothalamus, some is converted to estrogen by aromatase, an enzyme expressed by neurons in sexually dimorphic regions of the hypothalamus. If this enzyme is blocked, then the brains of male mice develop as female brains and the brains of female mice are resistant to the masculinizing effects of neonatal testosterone.[5] The effects of neonatal testosterone exposure are evident in the density of neurons in certain hypothalamic areas, in the connections that neurons in these areas make with other brain regions, and in the phenotypes of neurons in these areas.[6] The effects are exerted on neurons that, in early life, express receptors for estrogen; some of these neurons will die, others will change their phenotype and express a different complement of peptides and receptors in later life—and these changes will indirectly affect neurons to which they are connected. One important consequence is that, in adult life, the brain that had developed in the male pattern will respond differently to sex steroids than a brain that has developed in the female pattern.

In the 1980s, Nederlandse Hersenbank (the Netherlands Brain Bank), was set up by Dick Swaab, the head of the Netherlands Institute for Brain Research, to study the effects of aging and other factors on the human brain. Vasopressin was one of the few peptides that could be reliably identified in human postmortem brain tissue, and Swaab and his coworkers set about studying how vasopressin cells differed according to sex, age, and other factors. In the suprachiasmatic nucleus, there seemed to be many more vasopressin cells during the day than at night, and many more in summer than in autumn.[7] The cells themselves couldn't be disappearing and reappearing— there had to be marked changes in how much vasopressin they were producing. The shape of the nucleus also differed between males and females, although the number of vasopressin cells was similar between sexes: the size of the vasopressin cells was changing.

At this time, AIDS was scything down young lives, especially those of homosexual men, and some of the brains of this cruel harvest were donated to the Brain Bank. When Swaab and his team looked at *these* brains, they found that the suprachiasmatic nucleus contained, on average, more than twice as many vasopressin cells in homosexual men who had died from AIDS as in heterosexual men who had died from AIDS.[8] Knowing that male rats exposed prenatally to an aromatase inhibitor show "bisexual" partner preference, Swaab and colleagues went on to look at the brains of these rats—and they, too, had many more vasopressin cells in the suprachiasmatic nucleus than control rats.[9]

Steroid hormones, thyroid hormones, and prostaglandins all have diverse actions on the brain as well as on peripheral tissues. But most hormones are peptides: the anterior pituitary gland secretes prolactin—which (among many other things) regulates milk production—and five other peptide hormones, and is regulated by at least a dozen peptide hormones secreted by neuroendocrine neurons. From the anterior pituitary, thyroid-stimulating hormone regulates the production of thyroid hormones, and is regulated by thyrotropin-releasing hormone; LH and FSH regulate the gonads and are regulated by GnRH; ACTH regulates the adrenal cortex and is regulated by CRH and vasopressin; and growth hormone, regulated by GHRH and by somatostatin, regulates muscle and bone growth by regulating the secretion of insulin-like growth factor-1 from the liver. α-MSH from the intermediate lobe of the pituitary regulates melanin production. Insulin, amylin, and glucagon from the pancreas control glucose homeostasis. Secretin, the first hormone to be discovered, is secreted from the small intestine and regulates gastric acid secretion and bicarbonate homeostasis. Parathyroid hormone regulates calcium homeostasis. Relaxin from the ovaries controls labor, by softening the cervix. Angiotensin from the kidneys regulates blood pressure. Natriuretic peptides from the heart regulate sodium excretion. Cholecystokinin from the duodenum regulates digestion of fat and protein, and gastrin from the stomach regulates gastric motility. Other hormones from the gastrointestinal tract, including ghrelin, peptide YY, and glucagon-like peptide, signal to the brain to regulate energy balance, as do leptin and other adipokines secreted from adipose tissue.

This is a very partial list, and in many ways it is misleading. Each of these hormones has not one target but many, and it's often hard to be sure which role is most important, or whether that is even a sensible question. They act at multiple spatial scales, with autocrine effects on the cells that secrete them, paracrine effects on neighboring cells, and hormonal effects on distant targets. These actions are determined by the regulation of production and of secretion, and by the regulation of receptor expression in the target tissues. Their actions include acute activational effects on targets, slower effects on gene expression, and organizational effects on tissue development.

It is probably simplest to say that every cell in the body produces many different molecules that act both as feedback signals to the cells that produce them and as signals to other cells, sometimes only to close neighbors, sometimes to more distant cells. The diversity of signals is enormous, the receptors for them are even more diverse, and the ways that receptors are coupled to cellular function are more diverse again.

The concept of a hormone arose from the discovery by Bayliss and Starling in 1902 of secretin, a factor secreted from the duodenum that regulates pancreatic secretion. Three years later, Starling wrote: "We recognized these so-called internal secretions were merely isolated examples of a great system of correlations of the activities, chemical and otherwise, of different organs, not by the central nervous system but by the intermediation of the blood by the discharge into the blood stream of drug-like substances in minute proportions which evoked an appropriate reaction in distant parts of the body."[10]

A hormone, as first understood, was characterized as a substance produced at one site that was secreted in the blood to act at another, and which exerted effects in proportion to its concentration at the target tissue. When defined in this way, hormones are radically different from neurotransmitters. A neurotransmitter carries a message from one cell to another at a specific place—a synapse—and at a specific time—its effect lasts just a few milliseconds. These messages are targeted by the anatomical arrangement of interneuronal connections. By contrast, a hormone carries a message from one population of cells to another; its effects are extended in time, and depend on the ability of target cells to specifically, selectively, and proportionately recognize its presence.

Neurotransmitters are released one vesicle at a time, as discrete signals. When a vesicle is released into one side of the synaptic cleft, the concentration of neurotransmitter in the cleft reaches high levels, and some neurotransmitter will bind to receptors on the other side of the cleft. These receptors are molecules that span the cell membrane; on the extracellular side, the receptor molecule forms a "binding pocket" into which only molecules of a certain shape can fit. Molecules that fit in this way are called *ligands* for the receptor, and receptors are characterized in terms of their *affinity* for those ligands. Affinity is a measure of the probability that a molecule of ligand in solution will bind to the receptor. It takes only one molecule to activate a receptor molecule, but receptors with a low affinity need to be exposed to a high concentration of ligand for it to be likely that a molecule will bind to them; those with a high affinity can be sensitive to very low concentrations. Receptors for neurotransmitters do not need a high affinity for their ligands; on the contrary, it is perhaps better that they have a low affinity to ensure that the action is brief and local. The brevity of the actions of neurotransmitters means that they can transmit signals that vary rapidly over time.

Hormones are, by contrast, continuous signals, with half-lives of several minutes in the blood. Their effects are proportional to concentration and are mediated by receptors with very high affinity. They act selectively on populations of cells that express the relevant receptors, and those receptors are specific: they are not activated by any other signals unless present at very high concentrations. These characteristics—specificity, selectivity, and proportionality—are conferred by the density of expression of receptors.

Endocrine cells have neither dendrites nor axons, but many are like neurons in other ways. Some are electrically excitable: when pancreatic beta cells see an increase in extracellular glucose concentration they fire in bursts of spikes that are like the phasic bursts of vasopressin neurons; these bursts lead to calcium entry and trigger insulin secretion. In both neurons and endocrine cells, peptides are packaged in vesicles just as neurotransmitters are. Typically, peptide secretion is the result of the same process as that by which neurotransmitters are released: exocytosis is triggered in both cases by an increase in intracellular calcium. In neurons, this happens when spikes depolarize the neuron, opening voltage-sensitive calcium channels, and the same occurs in spiking endocrine cells.

However, endocrine cells have another trick. The cell bodies of all eukaryotic cells contain rough endoplasmic reticulum, which sequesters free calcium, and activation of receptors for some neurotransmitters or hormones can release calcium from these stores. In many endocrine cells, this "calcium mobilization" can trigger exocytosis of vesicles without any involvement of spikes. There is no rough endoplasmic reticulum in axon terminals, so spikes are necessarily involved in the release of synaptic vesicles.

Although the mechanisms by which vesicles are released are similar for synaptic vesicles and the vesicles that contain peptides, there are differences. Synaptic vesicles are only found at synapses, which contain specific release sites where just a few of the vesicles are "docked" ready to be released. A spike that invades a synapse will seldom release more than one vesicle, and often will release none; this is a stochastic process, and is erratic. The strength of a synapse can be quantified by the probability that a single spike will release any neurotransmitter, and this can change dynamically. The release probability depends on the size of the readily releasable pool—how many vesicles are docked—and declines as the pool is depleted. Conversely, when two spikes occur close together, the release probability can be higher for the second spike than for the first.[11] But endocrine cells have no specialized release sites; exocytosis can occur anywhere on the cell membrane. Vesicles are not generally waiting to be released, but, in response to signals, they are delivered to a readily releasable pool from stores deep inside the cell. This "trafficking" is organized by the cytoskeleton, an intracellular scaffold of actin fibers and microtubules that can either impede or facilitate access of vesicles to the plasma membrane.[12]

Magnocellular oxytocin and vasopressin neurons combine properties of classical neurons and classic endocrine cells. Like classical neurons, they receive inputs at their dendrites, generate spikes at the cell bodies, and propagate these spikes down their axons to trigger release from the nerve endings. However, their dendrites are also packed with vesicles, and these can be released by signals that mobilize intracellular calcium.

Some peptides can trigger secretion from the dendrites while inhibiting electrical activity. α-MSH acts at MC4 receptors on the oxytocin cells to mobilize intracellular calcium; this triggers oxytocin release from the dendrites, but also increases the production of endocannabinoids, which inhibit spiking activity. Thus α-MSH suppresses oxytocin secretion from the pituitary, which depends on spike activity, but stimulates oxytocin release within the brain. By combining properties of neurons and endocrine cells, oxytocin cells can independently regulate what they release centrally and peripherally.

This is exciting but unfortunate. We might wish, as many have wished, that by measuring oxytocin in the blood we might infer what is happening within the brain. We cannot. We might have hoped, as many have hoped, that by recording the electrical activity of oxytocin cells we might know what they are releasing within the brain. We cannot.

These characteristics may be common to the many types of neuron that make peptides in abundance. Vasopressin release from dendrites is even more abundant than oxytocin release, and is similarly governed by intracellular calcium mobilization. Generally, the large vesicles in which all secreted peptides are packaged are not found only or even mainly at synapses. They are distributed throughout the cytoplasm, in dendrites, cell bodies, axons, and terminals, and it appears that for peptides generally there are no specific release sites: exocytosis can occur from any compartment. The textbook image of a neuron is misleading: typically dendrites are shown as minor appendages to the cell, but they normally constitute about 85% of the cell volume. Dendritic release of peptides may not be an exception but the norm.

Endocannabinoids too are widely distributed in the brain; they, like steroids, are not stored in vesicles but are produced on demand. They have many effects, including on appetite, sleep, stress responses, and anxiety-linked behaviors, and the wide distribution of endocannabinoids and their receptors suggests that they are a common retrograde signaling mechanism. There are many others—the gas nitric oxide is one other, produced in some neurons by the enzyme nitric oxide synthase, but also in the endothelial cells of blood vessels by a slightly different version of the same enzyme. Another is adenosine; it is produced by most neurons when they are active and exported by a membrane transport mechanism to the extracellular fluid, where it can act at adenosine receptors on neighboring cells to inhibit them: caffeine has its stimulatory effects on brain function by blocking some of these receptors.

We have classically thought in terms of hierarchies, of higher centers, master controllers and slaves, systems and subsystems, placing the cerebral cortex at the top as the organ of personal responsibility. But all parts of the brain are massively and reciprocally interconnected. The organization is parallel, not serial; information flows between structures equally in two directions, and does so at every level of organization—between neurons and their "inputs," between neighboring neurons, between cells within networks, and between networks.

We can go further: the brain is massively and reciprocally connected with all parts of the body, the organization is parallel, not serial, and information flows between brain and organs in both directions and between organs and other organs. The brain communicates with organs by both neuronal signals and endocrine signals, and peripheral organs communicate with the brain by both neuronal signals and endocrine signals. The gut has an extensive nervous system all of its own, the enteric nervous system: in humans this contains about *500 million* neurons. The brain does far more than we are conscious of, and what we are conscious of is molded by a multitude of things that we are not conscious of.

21 Behavior

The soft thing looked askance through the window: he possessed the power to depart, as much as a cat possesses the power to leave a mouse half killed, or a bird half eaten. Ah, I thought, there will be no saving him: he's doomed, and flies to his fate! And so it was: he turned abruptly, hastened into the house again, shut the door behind him; and when I went in a while after to inform them that Earnshaw had come home rabid drunk, ready to pull the whole place about our ears (his ordinary frame of mind in that condition), I saw the quarrel had merely effected a closer intimacy—had broken the outworks of youthful timidity, and enabled them to forsake the disguise of friendship, and confess themselves lovers.

—Emily Brontë (1818–1848)[1]

This passage from *Wuthering Heights* seems to capture much of what neuroendocrinologists mean when they talk about behavior. It has little to do with reason and rationality, everything to do with emotions and instincts, as epitomized by Emily Brontë in the characters of Cathy and Heathcliff.

Neuropeptides are instruments of our emotions, and triggers for instinctive behaviors. In the last twenty years, the role of oxytocin in romantic love has caught the popular imagination.[2] In the early days, I gave a talk related to this on the afternoon of the last day, a Friday, of the annual meeting of the British Association for the Advancement of Science, in a session that included talks on the treatment of acute brain injuries. The meeting attracted a lot of media attention, but this session came too late in the program for the deadlines of daily newspapers, and the organizers had not anticipated any interest in my talk, shadowed as it was by matters of such profound substance.

As the week progressed toward that talk, I began to get urgent requests. Would I be available for a press conference on the Tuesday? Wednesday? Thursday? One was finally scheduled for the Friday morning, when all five speakers could attend. We agreed to speak for two minutes each before inviting questions, and I was last to speak. The two minutes of the other speakers lasted four or five, and as time passed I cut mine down. Mine was the first question: "Could I finish my sentence, please?" The room laughed, and I sat back as questions rained down on the others. Afterward the organizers reassured me—they were sure my talk would be very interesting and I shouldn't be troubled that the press didn't think so.

When I left the room, after helping to rearrange the chairs, the press were waiting, a posse in ambush. As each in turn fired a question, the phalanx lunged toward me with outstretched recorders, only to withdraw with a collective groan, as I parried, refusing to be drawn on any question about love. My talk would be about bonding in animal species, especially prairie voles; but I had never worked on voles or bonding. My contribution, with Mike Ludwig and others, had been to show part of the likely mechanism by which this behavioral change emerged. Love was something for poets to talk about; I had no insight beyond that open to every one of us through our own experience. After a dozen questions and a dozen groans, one journalist snapped—"Are you saying that this animal work is of no relevance to people?" Goaded, I replied, "Prairie voles, when they first meet, have sex more or less continuously for about thirty-six hours—would you be surprised if something happened in your brain?"

The press seemed to evaporate. The next day, I was the subject of editorials and cartoons and seemed to have given exclusive interviews to newspapers from Finland to America. A week later, a couple wrote to the *Glasgow Herald* to state that they had taken my advice, and finding it helpful, had adopted a vole in my honor at the Glasgow Zoo. None of the journalists had attended my talk, but they didn't misquote me; none had been unkind. It was frustrating, because what they were so interested in was, to me, trivial—*obviously* things happen in our brains when we have sex and fall in love. We had put a name to a small part of what happens. But the exciting thing was that a messenger released by neurons can act like a hormone within the brain, with profound and long-lasting effects on behavior; we had glimpsed how this might happen, and what we had glimpsed didn't fit comfortably with conventional views of the brain.

Oxytocin in the brain acts like a hormone, it acts at sites distant from where it is released, and it has long-lasting "organizational" effects. It has many effects on behavior: it suppresses appetite, especially for sweet, carbohydrate-rich foods; it stimulates male sexual arousal and female sexual receptivity; it promotes maternal behavior; it can suppress stress responses and reduce anxiety. Many of the brain regions involved, including the amygdala, the ventromedial nucleus, and the olfactory bulbs, are virtually devoid of oxytocin-containing axons but densely express oxytocin receptors. Thus oxytocin reaches some of its targets not by conventional synaptic release, but by "volume transmission," the flow of oxytocin released from dendrites through the extracellular channels of the brain.

After a rat gives birth, she displays a complex repertoire of maternal behaviors. If given paper, she will shred it and use the strips to build a nest. Virgin and early pregnant rats avoid newborn pups, but a mother rat will gather her young into the nest and allow them to suckle, and if any pup wanders away she will promptly retrieve it. Indeed, she will retrieve any pup that she sees close to her nest, whether hers or not, seemingly without limit. If 20 or 30 strange pups are placed in her cage, all will be retrieved, and she will strive to ensure that all are groomed and fed. She will be aggressive in defense of her nest and young, and will attack any intruder using a distinctive combination of lunges and bites. These behaviors are expressed after normal vaginal delivery, but are disrupted by interventions that impair oxytocin release, and are absent if the pups are delivered by caesarean section. Strikingly, maternal behavior can be induced by injecting small amounts of oxytocin into the brain, both in virgin rats when at the stage of the cycle when estrogen levels are high, and in ovariectomized rats that have been infused with estrogen.[3]

Lactating mice will retrieve mouse pups placed in their cage in response to the "distress calls" of the pups. Virgin mice do not usually show this behavior, but will do so if given an injection of oxytocin into the brain or if oxytocin neurons of the paraventricular nucleus are activated optogenetically. In this case, it seems that neurons of the auditory cortex that express oxytocin receptors are involved; only a few oxytocin-containing fibers enter this region but these seem to be enough to induce a prolonged enhancement of the salience of pup distress calls.[4]

While rats lavish maternal care on the offspring of other rats as freely as on their own, sheep are more discriminating. A ewe recognizes her own lamb from the moment it is born, and she nurtures it alone. But although her maternal behavior is unlike that of rats, oxytocin is similarly important. As the lamb, in being born, passes through the cervix and vagina, oxytocin is released into the mother's brain, and some reaches the olfactory bulbs.[5] A ewe recognizes her own lamb by smell: this "olfactory memory" is "fixed" by the oxytocin that reaches the bulbs.[6]

Managing flocks of sheep presents sad dilemmas; sometimes a ewe will give birth to twins or triplets but produce only enough milk for one to survive. Sometimes a ewe dies giving birth, and sometimes a lamb is stillborn, and the shepherd might want to foster an orphan with a ewe whose own lamb has died. In such cases, manual stimulation of the ewe's vagina and cervix can be used to induce a ewe to accept a strange lamb as her own.

Like rats, we humans are social animals. The bonds we make are profoundly important to us. In her last years I talked to my mother about her childhood, about how poverty had affected her life. She would have none of it: "We were rich," she insisted, speaking of the eight children living in four rooms with their widowed mother. "We were surrounded by love. And everyone helped everyone else."

Sociability is not about indiscriminately displaying friendliness; for us as for rats it is highly discriminatory. Few of us feel comfortable among strangers with whom we are forced to interact. We interact comfortably only with those we have come to trust: our family and friends, and, more circumspectly, with those we recognize as sharing key characteristics—by their accents, age, dress, and demeanor. But we would never find a mate if we were paralyzed by apprehension of strangers. Overcoming social apprehension is an important part of sexual behavior, and the ultimate conquest of that apprehension occurs when we form a bond with another person.

Rats are cautious. They avoid brightly lit or open areas of an unfamiliar cage, preferring to stay close to the walls or in covered areas, but in time will become confident in their familiar territory. They are sociable, preferring to be housed with other rats that they learn to recognize by smell. A strange juvenile rat placed in the cage of an adult rat will not be attacked, but will be systematically sniffed. If the juvenile is removed and returned later to the cage, it will be recognized from its remembered smell as familiar and unthreatening. Strange adult rats are a potential threat, especially to a lactating rat, and to challenge an intruder a lactating rat must shed normal caution. Oxytocin makes rats braver, and this effect is exerted in the amygdala.

If a rat is placed for a few minutes each day in a box, where it is given mild but irregular electrical shocks, after a few days it will "freeze" as soon as it is placed in the box. The role of oxytocin in this behavior was revealed by Valery Grinevich and his colleagues in experiments using optogenetics. They introduced channelrhodopsin2 into oxytocin cells of rats to make them sensitive to blue light, and trained the rats to expect a mild shock when placed in a box.[7] When blue light was shone onto the amygdala through a tiny implanted probe, the freezing behavior vanished. Thus oxytocin neurons, by their actions on the amygdala, reduce fearfulness, and this seems likely to contribute to the courage that lactating rats show when confronting an intruder.

The amygdala is close to the supraoptic nucleus and contains abundant oxytocin receptors. There are few oxytocin-containing fibers in the amygdala; those that are present look more like dendrites than axons, and they come mostly from magnocellular neurons. Valery and his colleagues also recorded from the amygdala in brain slices, and studied how neurons there responded to optogenetic stimulation of these few oxytocin fibers. The stimulation activated a subset of inhibitory GABA neurons. This activation was mediated in part by oxytocin, with slow and prolonged effects, but also in part by glutamate, which had rapid, short-lasting effects. Oxytocin cells also use glutamate as a conventional neurotransmitter: it seems that all peptide-containing neurons in the brain are "bilingual"—they all use a conventional neurotransmitter when speaking "privately" to some neurons, and the peptide when broadcasting a message of general importance to many.

Humans are one of a very few mammalian species that form monogamous bonds. We might think that the ability to love is one of those things that makes us human, and surely it is. Without love we would have no poetry, no song, no literature of note. Perhaps even language evolved not for us to discuss the weather but through sexual selection, for finding a fit partner and forming, honing, and keeping the bond that is so important for nurturing and sustaining our young. One of the rare mammalian species that, like humans, is monogamous and highly parental is the North American prairie vole *Microtus ochrogaster*. Even in the laboratory, adult prairie voles usually sit side by side, and the offspring make frequent ultrasonic calls and give other indications of distress when isolated. By contrast, the closely related montane vole is polygamous, minimally parental, and avoids contact with conspecifics; its offspring seem indifferent to social isolation.

A remarkable paper published by Tom Insel in 1992 was prompted by the idea that oxytocin might be involved not only in maternal bonding and maternal behavior but also in other social behaviors.[8] Insel compared the brains of montane voles and prairie voles: he found no differences between these species in the oxytocin cells, but speculated that there might be a difference in where oxytocin acted. At that time it was not possible to map the oxytocin receptors, but it was possible to see where they were likely to be by using autoradiography. This involves cutting thin sections of brain tissue and exposing them to radioactively labeled oxytocin. The sections are washed, leaving only the labeled oxytocin that is bound to oxytocin receptors, and then placed next to photographic film, resulting in images of the brain that display the density of binding. In the prairie vole brain, Insel saw dense binding in the bed nucleus of the stria terminalis, which is richly innervated by oxytocin-containing axons, but also in the lateral amygdala, the cingulate cortex, and the midline thalamus, which have few if any fibers. In the montane vole, there was little binding in any of these areas.

Did this pattern correlate with any aspects of social behavior? Insel noted that the montane vole becomes parental after parturition, and the level of oxytocin binding in the lateral amygdala increased after parturition almost to the level observed in the prairie vole. Then, with Sue Carter and others, he reported now-famous experiments demonstrating that the formation of monogamous bonds in the female prairie vole depends on oxytocin.[9]

It had long been suspected that oxytocin was released in women during sexual activity. The first to suggest this was probably Geoffrey Harris, who had been intrigued by the observation that oxytocin caused uterine contractions in the empty, nonpregnant uterus and had speculated that coitus might trigger oxytocin secretion to facilitate the transport of seminal fluid up the female reproductive tract. He predicted that vaginocervical stimulation would cause oxytocin to be secreted during coitus, and that, in lactating women, this would produce milk letdown.

He and Vernon Pickles found a novel way of testing this—they asked the wives of their colleagues, and they published the outcome in a letter to *Nature*, the title of which gives no hint of its unusual content.[10] In fact, they did not themselves ask the wives, but persuaded a new PhD student to do so, as Barry Cross, who was that PhD student, told me. Six wives had noticed milk letdown during coitus, though not necessarily at orgasm, and two more reported a "tingling experience" in their breasts like that experienced during suckling. Because oxytocin is essential for milk letdown, this "bioassay" was good evidence that oxytocin is secreted.

Whether peripheral secretion of oxytocin during sexual activity has any physiological role is not known: it seems to have little if any role in sperm transport, as once thought possible. Oxytocin is secreted during coitus in female goats, but only inconsistently in rabbits. In ewes, oxytocin secretion increases in the presence of a ram, but does not rise further during mating. Probably more important than secretion into the blood is the release of oxytocin into the brain that accompanies stimulation of the vagina and cervix.

When prairie voles first meet and mate, they engage in repeated intense sexual activity for a day or two, and this is critical for forming a pair bond. The bond can be revealed by a "partner-preference test": this involves separating the male and female for a few days, and then giving the female the choice of whether to spend her time with the male with whom she has mated or a new male. Montane voles don't care, but prairie voles usually choose the male with whom they had previously mated. Insel and his colleagues used this test to show that female prairie voles don't show any partner preference if given an oxytocin antagonist into the brain before mating. Conversely, virgin prairie voles who have met a male, but not been allowed to mate with him, will form a partner preference if given oxytocin into the brain at the time of first meeting.[9]

One obvious question was, What about the boys? Males bond with females just as females bond with males, but they show different behaviors. Perhaps the most consistent sign of bonding is that the male tends to express aggression toward other males—what we might think of as jealousy, or territorial behavior.[11] In several species, male bonding behaviors have been linked not to oxytocin but to *vasopressin*, including social memory in rats,[12] scent marking and aggressive behavior in hamsters,[13] and partner preference formation and paternal behavior in monogamous rodents.[14] These effects are linked to the distribution of V1a receptors in the brain, which, like that of oxytocin receptors, differs greatly between species and in consistent ways between monogamous and nonmonogamous species of rodents. In prairie voles, but not in promiscuous montane voles, male behaviors associated with monogamy are facilitated by central injections of vasopressin and can be prevented by a vasopressin antagonist. In monogamous rodents, V1a receptors are densely expressed in a part of the ventral forebrain that is involved in behavioral reinforcement. If the expression of V1a receptors here is further increased, by using a viral vector to deliver an additional gene load, male voles show increased anxiety and affiliative behavior, and they form strong partner preferences after overnight cohabitation with a female even when not allowed to mate with her.[15]

Remarkably, even transgenic mice engineered to express the prairie vole version of the V1a receptor show increased affiliative behavior when injected with vasopressin.[16] This was an extraordinary finding: it seems that a minor difference in the regulatory region of the gene for a peptide receptor could change the social behavior of an animal. In humans, the V1a receptor gene and the oxytocin receptor gene are polymorphic; variants are common in people, and there is some evidence that these differences predict individual variation in social attachment behaviors.[17]

The realization that the brain used so many different kinds of chemicals, in addition to classical neurotransmitters, to communicate between neurons was just the first step in a major conceptual shift in neuroscience. Many of these substances are neuropeptides, and most of those affect mood and behavior. The specificity of their effects resides not in the anatomical connectivity between neurons, but in the distribution of receptors within the brain. Different receptors have very different patterns of distribution, and the distributions differ between species in ways that correlate with differences in behavior.

That receptors for a peptide are densely expressed in some regions of the brain where there are few fibers containing that peptide is not a circumstance peculiar to oxytocin. In 1985, when the study of peptides in the brain was still in its infancy, Miles Herkenham drew attention to how frequent such mismatches were.[18] His catalog of mismatched neuropeptides included tachykinins, neurotensin, somatostatin, cholecystokinin, CRH, VIP, thyrotropin-releasing hormone, calcitonin gene-related peptide, angiotensin II, atrial natriuretic peptide, opioid peptides, and NPY. Of the neuropeptides he listed, tachykinins have been linked to nociceptive behavior,[19] neurotensin to vocal communication and social behaviors in starlings,[20] somatostatin and CRH to active and passive fear responses,[21] thyrotropin-releasing hormone to food reward,[22] angiotensin to cardiovascular reactivity and anxiety-like behavior,[23] CGRP to pain responses,[24] cholecystokinin to panic symptoms,[25] and atrial natriuretic peptide to thirst and impulsivity[26] and hence to alcohol intake;[27] opioids that signal through the mu opioid receptor promote anger and pleasure while inhibiting fear and sadness in humans.[28] NPY is very important for feeding, as explained previously, and VIP for organizing circadian rhythms.

The mere fact of a receptor-peptide mismatch in a particular brain area might have no great importance. It might be that many cells are promiscuous in the receptors that they express: if some receptors see no ligand, the cost to the cells is negligible. Profligate receptor expression might contribute to the evolvability of neural systems, and might be common because organisms with a liberal attitude to receptor expression are those most likely to acquire novel functions. Because extrasynaptic signaling does not require precise point-to-point connectivity, it is intrinsically "evolvable": a minor mutation in the regulatory region of a peptide receptor gene, by altering the expression pattern, could have functional consequences without any need for anatomical rewiring.[29]

That peptide receptors have distinctive patterns of expression, and that peptides produce coherent behavioral effects when given quite crudely into the brain, suggest that volume transmission is used as a signaling mechanism by many different populations of peptidergic neurons. We thus must see neuropeptides as "hormones of the brain."[30] Once we begin to think in these terms, we are led to ask in what way do the hormones that come from within the brain differ from the hormones from peripheral organs that act on the brain?

Some hormones, like cholecystokinin, act on afferent endings of the vagus nerve, and their messages are relayed to the brain by classical neuronal pathways. Some, like leptin from adipocytes, insulin from the pancreas, and prolactin from the pituitary, enter the brain by specific transport systems. Yet others act on the brain at specialized sites where there is no blood-brain barrier at all. Three of these sites, the subfornical organ, the subcommissural organ, and the organum vasculosum of the lamina terminalis, are at the front of the hypothalamus; these curious structures dangle into the third ventricle and are full of "leaky" blood vessels. Another such site, the area postrema, faces the fourth ventricle and is densely interconnected with the dorsal vagal complex, so important in the regulation of the gut and of appetite. Each of these "circumventricular organs" lacks a blood-brain barrier, and neurons in each express a wide variety of peptide receptors and can respond to many hormones that do not enter the brain proper. The median eminence, adjacent to the arcuate nucleus, also has a leaky blood-brain barrier, and some blood-borne peptides affect the activity of arcuate neurons by penetrating the brain at that site.

One behavior affected by blood-borne hormones acting at circumventricular organs is salt appetite. As omnivores, we never experience sodium deprivation unless we drink excessive amounts of water and actively exclude sodium from our diet, but many mammals barely get enough salt from their diet, and when deprived of sodium they will avidly seek out salt and will choose to drink saline in preference to sweet solutions.[31] This behavior is triggered by the hormone angiotensin: when sodium concentrations are low, the kidneys secrete renin, which is converted in the blood to angiotensin. Angiotensin acts on the subfornical organ, and the salt appetite that it evokes involves two other neuropeptide systems—orexin neurons in the lateral hypothalamus, and neurons in the caudal brainstem that express relaxin.[32]

The Selfish Gene (1976), by Richard Dawkins, made a lasting impression on me as on many others, not because it is well written, though it is, nor from what I learned from it, though I learned much, nor because its fundamental thesis is right, which I doubt, but because it turned an idea on its head and made us think afresh. Dawkins challenged us to think not of genes as ways to make bodies, but of bodies as ways to make genes. We might similarly question how we think about the brain: from the perspective of an adipocyte, for example, the brain is just something that follows its instructions to keep it supplied with lipid.

Communication between our brain and the other organs of our bodies goes in both directions, and just as our brains control how our organs behave, so they control how we behave, and they do so through the *heart of the brain*, the hypothalamus. We might like to believe that how we act, what we think, and the choices we make are governed by our reason. Biology tells us otherwise.

The "hormonal" nature of peptide signaling in the brain suggests that certain peptide signals, such as oxytocin, convey *emotional salience*; that they might affect behavioral decisions by differentially weighting sensory information according to its relevance to emotional "needs." If this is the case, it seems credible that, just as interventions through drugs that have widespread effects on signaling through dopamine, noradrenaline, and serotonin have pronounced, sustained, and reproducible effects on mood and on specific behaviors, similar interventions that alter the levels of particular neuropeptides might be expected to have specific effects. In recent years, considerable attention has been given to interventions in humans aimed at increasing the levels of oxytocin in the brain. In particular, a great many studies have examined the effects of administering oxytocin intranasally, and many of these have reported effects broadly consistent with the idea that oxytocin modulates behaviors that depend on awareness of "social" cues. These results have driven the notion that therapies based on oxytocin's actions might be useful in disorders of social functioning, such as autism.[33]

Mike Ludwig and I have been outspokenly critical of many of these studies.[34] We have noted that the amounts of oxytocin delivered in intranasal studies are enormous, yet because of a very efficient blood-brain barrier only tiny amounts actually reach the brain. We have expressed concern that many of the reported effects might be mediated not in the brain but by actions at peripheral targets, and that controls for peripheral effects are generally absent in such studies. Others have questioned the replicability of effects of intranasal oxytocin on behaviors.[35] Others again have pointed out that the effect sizes reported are small and that studies have been consistently underpowered, and have suggested that publication bias and confirmation bias might be prevalent in this literature.[36] Studies are *underpowered* when the sample sizes are too small to reliably detect an effect of a given size. Why this is a problem can be seen as follows: if a study design is underpowered, and the same design is followed in five studies, perhaps only one of these will report a significant effect. Unfortunately, it is likely that only the study showing a significant effect will be reported in the literature: this is what we mean by *publication bias*—studies with inconclusive or negative results tend to be just filed away and forgotten. Finding a significant effect does *not* establish that the effect is real, only that that effect would not often arise by chance alone from the inherent variability of samples, so if a large number of small, underpowered studies are performed, then some will show results that are "false positives," and these will be reported and emphasized, while negative or inconclusive results go unreported, or if reported are discounted. This problem is confounded by *hidden multiple comparisons*. This issue can arise in many ways, including when a study does *not* produce the predicted outcome, but when subsets of the data are subsequently selected that do display an apparently significant effect. This is further confounded by *confirmation biases*—when the data that are presented, and the ways in which they are analyzed, are selected retrospectively to support the prior expectations of the experimenters. These are pervasive problems in the scientific literature, and not confined to studies of intranasal oxytocin. As John Ioannidis expressed it in the abstract to his seminal paper "Why Most Published Research Findings Are False":[37]

> The probability that a research claim is true may depend on study power and bias, the number of other studies on the same question, and, importantly, the ratio of true to no relationships among the relationships probed in each scientific field. In this framework, a research finding is less likely to be true when the studies conducted in a field are smaller; when effect sizes are smaller; when there is a greater number and lesser preselection of tested relationships; where there is greater flexibility in designs, definitions, outcomes, and analytical modes; when there is greater financial and other interest and prejudice; and when more teams are involved in a scientific field in chase of statistical significance…, for many current scientific fields, claimed research findings may often be simply accurate measures of the prevailing bias.

Studies of the behavioral effects of intranasally applied oxytocin have consistently involved delivering massive amounts of oxytocin, very little of which reaches the brain, but large amounts of which enter the peripheral circulation. The effects on social behavior have consistently been relatively small and poorly replicable. These have attracted a great deal of media attention and public interest, often driven by extravagant overinterpretation.

However, the *idea* that delivering oxytocin into the brain or enhancing oxytocin release in the brain might have coherent behavioral outcomes, and that this might be therapeutically beneficial in some cases, is credible. It is well supported by carefully controlled animal experiments, and the "shape" of a mechanistic understanding is at least dimly visible.

22 The Evolved Brain

The species of fossils, minerals, plants, animals, which are found in the Waters, and near the surface of the Earth, are still more intricately diversified; and if we regard the different manners of their production, their mutual influence in altering, destroying, supporting one another, the orders of their succession seem to admit of an almost infinite variety.... To introduce order and coherence into the mind's conception of this seeming chaos of dissimilar and disjointed appearances... it became necessary to suppose, first, That all the strange objects of which it consisted were made up out of a few, with which the mind was extremely familiar: and secondly, That all their qualities, operations, and rules of succession, were no more than different diversifications of those to which it had long been accustomed.

—Adam Smith (1723–1790)[1]

I open this chapter with a quote from an essay by Adam Smith, author of *The Wealth of Nations*. When it was written is not clear; it was published posthumously in 1795. Why Smith did not publish the essay in his lifetime is not known. He did not destroy it, as he destroyed most of his unpublished papers, but he might have worried about the theological implications. The parallels with his thinking on economics are as obvious as those with evolutionary theory, and he was a close friend of the geologist James Hutton, who was certainly wary of religious sensitivities. Hutton transformed our understanding of the earth by deciphering the messages carried by rocks. What he saw told that the geological features of our world have been formed over immense time by the continual operation of diverse forces, acting from above by continual erosion and from below by intense heat and pressure; that the present appearance of our world reflects processes that appear to have been continuing indefinitely.

The hypothalamus is sometimes described as primitive, but it might be better described as highly evolved. The hypothalamus and its hormones have their evolutionary roots in *Urbilateria,* wormlike marine organisms that are the last common ancestor of vertebrates, flies, and worms. In *Urbilateria,* peptide-secreting cells probably responded to cues from the ancient marine environment. In the mammalian hypothalamus too, many neurons respond to environmental signals: some to osmotic pressure, some to sodium, some to glucose, some to fatty acids and other nutrients, and some to temperature.

Oxytocin is a sequence of nine amino acids and it differs in only two of these from vasopressin. Other similar peptides are present in many invertebrates, including locupressin in insects, conopressin in gastropods, cephalotocin in cephalopods, and annetocin in annelids. The original ancestor of these peptides appears to have emerged at about the time of evolution of a symmetrical body plan. Generally invertebrates have just one such peptide. The only known exceptions are the octopus and the cuttlefish, which have both an oxytocin-like peptide (octopressin) and a vasopressin-like peptide (cephalotocin).[2] In the octopus, octopressin is involved in reproduction, cardiac circulation, and feeding: it can induce contractions in muscles of the penis, oviduct, rectum, and anterior aorta, tissues that are innervated by octopressin neurons. It also regulates the electrolyte composition of the hemolymph (the invertebrate equivalent of blood) and urine. Cephalotocin has none of these effects but is secreted into the hemolymph, so it might act as a circulating hormone.

Fish first appeared about 530 million years ago during the Cambrian explosion, an apparently sudden emergence of many diverse species with a profusion of body forms. The earliest fossils are of jawless fish, like modern lampreys. Lampreys have only one peptide that is like oxytocin and vasopressin, vasotocin, which differs from vasopressin by just one amino acid in its sequence of nine amino acids. The first jawed fish appeared about 440 million years ago and included cartilaginous fish and bony fish. All cartilaginous fishes have vasotocin and also an oxytocin-like peptide. The Pacific ratfish has oxytocin itself, but six other homologs have been identified in different species: aspargtocin, valitocin, asvatocin, phasvatocin, phasitocin, and glumitocin. Bony fish have vasotocin and an oxytocin-like peptide, isotocin; they evolved into ray-finned fish, which constitute nearly all of the more than 30,000 species of modern fishes, and into lobe-finned fish, living examples of which are coelacanths and lungfish. All four-limbed vertebrates (tetrapods), including amphibians, mammals, reptiles, and birds, evolved from lobe-finned fish and all have at least one homolog of oxytocin (usually isotocin, mesotocin, or oxytocin) and one of vasopressin (vasopressin or vasotocin).

It is not just the peptides that are highly conserved through evolution. In vertebrates, the peptide genes are also highly conserved in both their structure and sequence. The cDNA for conopressin in pond snails codes for a precursor molecule that is very like the oxytocin/vasopressin precursor, and the vasotocin genes from cyclostomes, teleosts, and chicken all have a three-exon structure like that of the vasopressin gene in mammals.

In ray-finned fishes, vasotocin is involved in osmoregulation and isotocin in regulating ionic concentration. It therefore seems that, in the aquatic vertebrates where vasotocin and isotocin first appeared as distinct hormones, both were involved in salt and water balance. They may also have been involved in reproduction, the timing of which is, generally, tied to environmental conditions. Isotocin also seems to have extensive effects within the fish brain: Gil Levkowitz at the Weizmann Institute in Israel estimates that of the 300 isotocin neurons in the brain of an adult zebrafish, only about 30 project to the pituitary—the rest innervate extensive areas of the brain and, as in mammals, modulate social behavior and fear responses.[3]

With the emergence of land vertebrates, isotocin evolved into oxytocin through an intermediary stage of mesotocin; mesotocin differs from isotocin by just one amino acid and oxytocin from mesotocin by another. Vasotocin evolved into vasopressin; again, this change involves just a single amino acid substitution. Land mammals secrete vasopressin to concentrate the urine, and in some, including rodents, oxytocin contributes to body fluid and electrolyte balance by promoting salt excretion. These observations in species separated by several hundred million years suggest that there has been strong conservation not only of the genes but also of some functions of oxytocin and vasopressin-like peptides.

Thus vasopressin and oxytocin arose through duplication of the vasotocin gene in a species of jawless fish that lived about 400 million years ago; a separate duplication event probably gave rise to octopressin and cephalotocin in octopus and cuttlefish. This "degenerate" duplication in jawless fish allowed the neuroendocrine systems to diverge, refining and elaborating distinctive roles. In the guppy, a viviparous teleost, isotocin is involved in the induction of parturition, so this role of oxytocin that we might have imagined to be particularly mammalian has an ancient origin.

While the duplication of a peptide gene is a necessary prerequisite for the evolution of two functional systems, much more needs explanation. For two descendants of a peptide to acquire differentiated functions, different receptors must arise through which their functions can be exercised, because the actions of peptides cannot be anatomically confined to a very localized region—once released, they persist in the extracellular fluid and can diffuse or be conveyed by the flows of that fluid, sometimes to distant sites. Secondly, the two peptides must be regulated differently, and this requires that they be housed in different cells.

Gene duplication is the main way in which new genes arise. When a gene is duplicated, one copy can usually maintain the original function, leaving the other free to mutate. The duplication is not necessarily harmful, because how much protein a gene will produce is regulated. Overproduction of a secreted product will generally be compensated for by downregulation of the relevant receptors. Organisms are thus very tolerant of different levels of gene expression, and after a gene is duplicated, mutations in one copy will not necessarily lead directly to any loss of function, even if the peptide product is overproduced. When a mutation prevents one copy from encoding a functional protein it may become a "pseudogene," a functionless remnant of the duplication event, prone to accumulating further mutations.[4] Occasionally, however, a mutation will be beneficial and one copy will acquire a new function and become a new gene. The human genome contains about 20,000 pseudogenes, and most of these probably have no function; they seem to be evolving without any selection constraints. These pseudogenes are evidence of an evolutionary history of extensive gene duplication and are part of the "junk DNA" that litters the human genome.

There are four known receptors for oxytocin and vasopressin. Vasopressin acts in the kidney tubule to regulate the water permeability of the renal collecting ducts at the V2 receptor. Other actions of vasopressin, in the brain, pituitary, and on blood vessels, are mediated by V1a and V1b receptors, while oxytocin acts on the mammary gland and uterus at oxytocin receptors. Orthologs of these four receptors (i.e., proteins in different species that are so similar in structure, sequence, and function that they can be considered to be the same) have been described in all vertebrates investigated so far.

The four receptors are very similar, suggesting that they originated from the same ancestral gene. During the Cambrian explosion there were two episodes of whole genome duplication in vertebrates, and it is probably these that gave rise to this set of receptors.[5] V2 receptors are different, and probably arose from an earlier gene duplication. Lampreys, which have only one member of the vasopressin/oxytocin family (vasotocin), have both V2-type receptors and V1-type receptors.[6] Thus, when the vasotocin gene was duplicated in early vertebrate evolution, there were already two families of receptors present that could allow the functions of descendant peptides to diverge.

For their functions to diverge, vasopressin and oxytocin also had to be expressed in different neurons. When a gene is initially duplicated, the two copies will normally be expressed in the same cells. Exactly where a gene is expressed in the body is determined by regulatory elements in the DNA that are usually close to the protein-coding region of the gene, so those elements are likely to be duplicated along with the protein-coding sequence. Identifying these regulatory elements in the rat genome is hard because of the large amount of junk DNA: the rat genome has 3 billion bases, but only a few of these are in protein-coding sequences. A small proportion of the genome encodes functional RNA molecules, and some sequences control gene expression and determine the structure of the chromosomes, but about 90% of the genome consists of repetitive, mutationally degraded material with no apparent function. This includes the pseudogenes, but much more common are transposable elements. These sequences have been called "parasitic" or "selfish" because of their capacity to multiply while (in most cases) serving no useful function.

The human genome contains a similar amount of junk DNA as does the rat genome. It includes about a million copies of one particular parasitic sequence of about 300 bases: these "Alu elements" comprise about a tenth of the genome and have been implicated in many hereditary diseases, including hemophilia and breast cancer.[7] By contrast, the pufferfish has just 390 million bases with very few repetitive elements, yet about as many genes as mammals (20,000–25,000). It might be imagined that the difference in size between the pufferfish and human genomes simply reflects the difference in complexity of the organisms. This doesn't hold up. While the human genome has 8 times more DNA than that of a pufferfish, it has 40 times less than that of a lungfish. More than 200 salamander genomes have been analyzed so far, and all are between 4 and 35 times *larger* than the human genome. The pufferfish is remarkable because, for reasons still obscure, it has managed to either shed junk DNA or avoid being encumbered by it in the first place.

In the fish preoptic area, isotocin and vasotocin are expressed in separate neurons. In Bristol, David Murphy and his colleagues produced transgenic rats by inserting 40,000 bases of pufferfish DNA that included the isotocin gene. In these rats isotocin was expressed only in oxytocin neurons, and, in response to dehydration, expression of both isotocin and oxytocin was stimulated in a similar way.[8]

From these experiments we can conclude a lot. Every cell type has its own molecular "password," a combination of a few genes that determine its identity. Genes with regulatory elements that recognize this password will then also be expressed in those cells. When a gene is duplicated, the two copies are normally expressed in the same cells because the regulatory elements that determine where it will be expressed are generally close to the gene and hence are also duplicated. Because the fish isotocin gene recognizes the mammalian oxytocin cell, that oxytocin cell must have the same password as isotocin cells in fish. This implies that the password arose early in vertebrate evolution and has been tightly conserved through subsequent evolution. Equally, the regulatory elements of oxytocin-like genes must also have appeared early in vertebrate evolution.

For a few cell types we can spell out the molecular password precisely. In zebrafish embryos, differential expression of two transcription factors, *nk2.1* and *pax6*, separates the developing forebrain into two. The *nk2.1+* region gives rise to the hypothalamus and preoptic area, and, within this, vasotocin neurons are determined by the expression of a tissue-specific microRNA, *miR-7*, and two other transcription factors, *rx3* and orthopedia (*otp*). Thus the molecular password (*miR-7+*, *nk2.1+*, *rx+*, *otp+*) defines the cell type (vasotocinergic extraocular photoreceptors), and the same password defines the same cell type in the invertebrate annelid worm *Platynereis dumerilii*. In the dorsal preoptic area of zebrafish, vasotocin cells mingle with cells that express isotocin, and both of these cell types project to the posterior pituitary gland. Two transcriptional regulators, *orthopedia b* and *simple-minded 1*, are required for expression of vasotocin and isotocin in this region.[9]

This vasotocin system is plastic and can be influenced by many factors, including behavioral factors such as social hierarchy.[10] In dominant individuals, the preoptic area has between one and three pairs of large vasotocin neurons, whereas subordinate individuals have between seven and eleven pairs of small vasotocin neurons in a slightly different location. In fish, vasotocin regulates electrolyte balance as vasopressin does in mammals and, again like vasopressin in mammals, it also influences social behavior. In the bluehead wrasse, a sex-changing fish with both territorial and nonterritorial males, vasotocin increases aggression in nonterritorial males but decreases it in territorial males.

So just as vasotocin in fish is homologous to vasopressin in mammals, the vasotocin cells of fish are homologous to the vasopressin cells of mammals, and are descended from the vasotocin cells of invertebrates. So how did the oxytocin cells diverge from vasopressin cells to become a distinct cell type?

It might be relevant that all magnocellular vasopressin cells also express some oxytocin, though mostly very little, while all magnocellular oxytocin cells express some vasopressin, though again mostly very little.[11] It's hard to believe that this coexpression is functionally meaningful. These neurons make massive amounts of their principal products: the machinery for synthesizing peptides is exceptionally active, because it must make enough to provide biologically effective concentrations in the systemic circulation. But very many other peptides are expressed at low levels in these neurons. Some of these "coexisting" peptides—like dynorphin—do have important roles, but why so many others are expressed at low amounts is not clear. Perhaps the cost of evolving ways of repressing such expression completely is not worth the modest cost of some promiscuous but biologically irrelevant expression. Not everything produced by a cell necessarily matters; gene expression is, like everything in a cell, noisy and messy.

However, a few neurons express both peptides at about the same, intermediate level, and the proportion that do so increases in conditions of sustained demand. Again it seems unlikely that this ambivalence has any functional significance, but it is nevertheless interesting that it should happen at all. Perhaps related to this is the observation that, although phasic firing is a characteristic of vasopressin cells, since the earliest studies of Wakerley and Lincoln we have known that a few phasic neurons display milk-ejection bursts during suckling. Thus, some neurons have phenotypic characteristics of both oxytocin cells and vasopressin cells. Are these the neurons that express intermediate amounts of both oxytocin and vasopressin? We don't know. These observations indicate that vasopressin cells and oxytocin cells have evolved ways to suppress the expression of the alternate peptide, and presumably, with that, to suppress the expression of things that determine the alternate neuronal phenotype. Oxytocin and vasopressin cells have many things in common but also have important differences: vasopressin cells express vasopressin receptors and oxytocin cells express oxytocin receptors, for example, and while both express dynorphin, vasopressin cells do so much more strongly. The channels that determine their intrinsic electrical properties are expressed in both clans, but at different average levels (though with considerable heterogeneity within each clan).

That a few neurons have an intermediate phenotype suggests that perhaps the developmental process of differentiation between oxytocin and vasopressin cells is itself a bistable dynamical system, that the fate of bipotential progenitor neurons is tipped one way or another by factors in the environment of the developing neurons. In other words, perhaps the same mechanism that induces some neurons to express high levels of oxytocin also represses the expression of vasopressin, and vice versa. By this reasoning the few ambivalent neurons might be a few that remain in an unstable equilibrium of cell fate determination. A similar circumstance seems to hold for neurons in the arcuate nucleus: all of the α-MSH cells appear to make some AgRP and NPY, but usually at very low levels, while those that make large amounts of NPY and AgRP all appear to produce some small amounts of α-MSH.[12]

The earliest neurons combined properties that we have thought of as separate properties of endocrine cells and neurons. They used a diversity of signaling mechanisms, made both peptides and neurotransmitters, and were endowed with a wide range of specialized senses. They had not a single role to which they were committed, but multiple behavioral and physiological functions. As these ancestor neurons proliferated in descendant species, populations differentiated not primarily by gaining properties but by losing properties. Paradoxically, we have become more intelligent as our neurons have become less complex.

Nevertheless, the neurons of the hypothalamus retain the multifunctionality of their distant ancestors, and their multitude of sensory abilities. Magnocellular oxytocin neurons regulate milk ejection, parturition, and sodium excretion by what they secrete into the blood, and there is no conflict between these roles: the uterus expresses abundant receptors to oxytocin only at term pregnancy, and the mammary gland only in lactation; the mammary gland "sees" only the pulses of secretion that occur at reflex milk ejection, while the kidney sees low concentrations, and what matters there is the secretion evoked by the background chattering of oxytocin neurons. The oxytocin neurons also govern reproductive and appetitive behaviors, and these are governed reciprocally, not by the oxytocin that is released into the blood but by oxytocin released from dendrites. Vasopressin neurons in the retina are sensitive to light; those in the supraoptic nucleus both to osmotic pressure and to temperature[13]—vasopressin is released in hot conditions to preserve body water in the face of evaporative loss. Both oxytocin and vasopressin neurons are sensitive to multiple chemical cues from the internal environment—they have receptors for glucocorticoids and gonadal steroids, and for leptin, prolactin, and insulin, as well as for many of the peptides released from the brain itself. Similar things are true of the other hypothalamic clans.

James Hutton was not a great writer, and his ideas gained influence only when re-presented by his friend the mathematician John Playfair. But he wrote one perfect line, summarizing his conclusion that the appearance of the earth's rocks suggested a long history of continual and continuing change: "The result, therefore, of our present enquiry is, that we find no vestige of a beginning,—no prospect of an end."[14]

Hutton's line encapsulates what we see when we peer into the human genome. We know there was a beginning, but every complex animal has a genome of similar complexity to all others. The genomes display evidence of continual change, but no evidence of progression. We have a brain a million times more complex than that of a zebrafish, but we do not have a million times as many genes; we have about the same number, and they are essentially the same genes. While we are good at being humans, zebrafish are good at being zebrafish: we are not more evolved, only differently evolved. So what makes us human, and more clever and resourceful than zebrafish if we believe ourselves to be so? We must look to understand our brains less as the product of new and better rules, but more as the product of repeated iteration of the same rules of development as are implemented in less complex animals. By this reasoning, intelligence and other higher functions are not the product of new genes and new mechanisms, but emergent phenomena that arise from complexity.

23 Redundancy and Degeneracy

If an organised body is not in the situation and circumstances best adapted to its sustenance and propagation, then, in conceiving an indefinite variety among the individuals of that species, we must be assured, that, on the one hand, those which depart most from the best adapted constitution, will be the most liable to perish, while, on the other hand, those organised bodies, which most approach to the best constitution for the present circumstances, will be best adapted to continue, in preserving themselves and multiplying the individuals of their race.
—James Hutton (1726–1797)[1]

How did the complex brains of vertebrates evolve? *C. elegans* has only about 300 neurons, each with a specific and unique function. To execute its function, each must exchange signals with other neurons and with the other cells whose function they regulate; they communicate using both electrical and chemical signals, including both neurotransmitters and hormone-like messengers. They receive signals from the animal's external and internal environment, and many have specialized sensory properties, enabling them to respond directly to stimuli such as light, osmotic pressure, electrolytes, nutrients, and a variety of chemicals. These neurons are multifunctional, combining properties of sensory cells, neurons, and endocrine cells.

In bigger brains, no single neuron is particularly important: each is part of a clan of hundreds or thousands of similar neurons that collectively execute a particular function. The zebrafish has a handful of vasotocin neurons; the rat many thousands of vasopressin neurons in several clans. The clans differ in many ways—in the shape and size of their neurons; in where they project to and where they receive inputs from; in what other proteins and peptides they make. Through the evolution of increasingly large and complex brains, the different neurons of simple brains have proliferated and diversified into distinct clans, each retaining some features of an ancestor cell while acquiring new features through mutations.

But there are important differences between a nervous system in which a function is enacted by a single cell and a nervous system in which the analogous function is executed by a clan of cells. The death of any one neuron in a simple nervous system has major consequences, but a single *C. elegans* lives for just two or three weeks, has a generation time of four days, and may produce a thousand progeny in each generation. We live for seventy years or more and raise just a few children; our lives are, in this sense, more precious. Our brains must be robust against the insults that our lives deliver and the random neuronal death that ensues with these insults, and the brains of our children must be robust against the vagaries of random gene mutation and a process of development that is, to a large extent, stochastic. That robustness is in part enacted by redundancy: when a function is enacted by more neurons than are needed, it can withstand a degree of attrition by random loss of neurons and tolerate inefficiency introduced by minor gene mutations or developmental imprecision.

Redundancy brings problems. A signal implemented by more neurons than are needed may be a signal that is too large, and a signal that must be precisely timed may be "smeared" if implemented by a large population of autonomous neurons. The clan response must be orchestrated; the members of the clan need to know what the other members are doing. The signals that pass among cells of a clan may have to be different from the signals that pass from a clan to its target cells: to efficiently inhibit a target the cells in a clan may have to excite each other, and we see examples of both mutual excitation and mutual inhibition in the hypothalamic clans.

There are other ways of implementing robustness. Redundancy, in the strict sense, is the overprovision of agents to fulfill a given task: this provides robustness against the death or inactivation of some of those agents but not against the inactivation of all of them. Robustness against that eventuality is provided when the same task is achieved by different means. The ability of structurally different elements to perform the same function is called *degeneracy*, and it is a characteristic of many biological systems. Degeneracy is both necessary for, and an inevitable outcome of, natural selection.[2] If a system is not degenerate then a mutation is likely to be lethal, but if it is degenerate then a mutation can open the door to the evolution of a novel function. The genetic code itself is degenerate: different nucleotide sequences can encode the same amino acid. Many genes are degenerate: transcription can start and stop at different sites and still yield the same protein product. At the highest level, our language is degenerate: we can express the same meaning in many different ways. In our language the existence of synonyms allows words to evolve in meaning over time: "viral" does not have the same meaning now that it did when I was young.

Degeneracy arises in biological systems at many levels and in many ways. Natural selection can only work in a population where there is heritable variability, and that variability comes from the accumulation of neutral gene mutations. A mutation may be neutral because it has no functional consequences, or because it degenerately or redundantly duplicates an existing function. Such mutations accumulate because there is no strong selection pressure to eliminate them. In the process of genetic recombination that accompanies sexual reproduction, new combinations of neutral mutations might display novel functionality, and, if adaptive, these will be selected for.

Our evolvability—our ability as a species to adapt to environmental change—thus depends on genetic variability and degeneracy. On a different timescale, our ability as individuals to adapt to life challenges also depends on degeneracy, in the form of our ability to solve problems by another way when one way is blocked. This poses problems for neuroscientists. We typically infer the function of a neuronal population from three sorts of experiments: by observing it in different physiological circumstances; by showing what effect its activation has on physiology or behavior; and by showing the physiological consequences of removing or disabling it. When these three sorts of experiments converge on a conclusion, we are likely to have reasonable confidence in that conclusion. But there are pitfalls.

If a neuronal population responds in some way to a particular physiological challenge, how do we know that that response has any physiological significance? Some responses may be too small or brief to be functionally relevant, while others may not be what they seem. At the end of pregnancy, oxytocin neurons respond to many stimuli less strongly than they do at other times. But at the end of pregnancy there is a massive store of oxytocin in the pituitary—the store has been upregulated in anticipation of the demands of parturition and lactation. Because the stores are greater, spikes release more oxytocin—so the weaker spiking activity seen in response to many stimuli evokes a normal amount of secretion. The attenuated spiking response is an adaptation that preserves a normal function.[3]

When we give a peptide into the brain, we typically inject a large amount to see a large effect, but if we give too much the peptide might act at *other* receptors. Oxytocin can act at vasopressin receptors and vasopressin at oxytocin receptors at concentrations only modestly higher than those at which the native ligand acts. If giving a peptide into a part of the brain changes some behavior, we need to know that in the behaving animal the endogenous peptide is actually released in relevant circumstances and reaches that site in sufficient amounts—because the receptors, as a result of a neutral mutation, might be expressed at sites that never see the endogenous ligand.

If we remove a peptide from an animal, we may see no effect because we have intervened in a degenerate system. This is not a hypothetical pitfall; any true and full account of how our understanding has emerged would be replete with such tales.

In the arcuate nucleus, one clan of neurons makes two peptides, NPY and AgRP. Both are potent orexigens: when injected into the brain in tiny amounts they stimulate eating. These neurons are primary drivers of appetite: they are activated when the empty stomach secretes ghrelin, which acts directly on the NPY/AgRP neurons, and are inhibited by leptin when our energy stores are in excess. The expression of mRNA for NPY and AgRP is increased after fasting, antagonists to either one of these peptides will inhibit feeding, and activation of these neurons by optogenetics will activate feeding. AgRP acts at the MC4 receptors through which the satiety signal α-MSH acts, but with the opposite effect of α-MSH—it is an "inverse agonist."

Yet transgenic mice that lack NPY have a normal body weight. Similarly, transgenic mice that lack AgRP have a normal body weight. So is the absence of either compensated for by the other? Transgenic mice that lack both NPY and AgRP also have a normal body weight. So did we get it wrong, are these neurons unimportant? To test this, two groups of scientists independently and at about the same time found ways to selectively kill the NPY/AgRP neurons. The outcomes were clear: when most of these neurons were killed, the mice stopped eating and were kept alive only by gastric feeding.[4,5]

It seemed that the neurons could still drive feeding despite the loss of both NPY and AgRP, and they could do so because they remained able to release the neurotransmitter GABA. Remarkably, delivering a GABA agonist to mice, using a tiny implanted pump to deliver it into the peritoneal cavity, completely prevented the anorexic effects of destroying the NPY/AgRP neurons—and the mice continued to eat normally even after stopping the infusion a few days later.[6] Thus, given just a few days, the mice could compensate fully for the loss of these neurons, at least as judged by their ability to eat a normal amount of food each day.

Some peptides are essential for the things we believe them to be important for: mice without GnRH or kisspeptin are infertile; mice without oxytocin cannot feed their young; rats without vasopressin cannot concentrate their urine; animals without growth hormone or its releasing hormone remain very small. Many, though, are not. Ghrelin, known as the "hunger hormone" for its potent effects on appetite and adiposity, was named not for this function but for its potency at stimulating growth hormone secretion. But ghrelin-deficient mice grow normally, eat normally, and have a normal body weight.[7]

Oxytocin is named from the Greek for "quick birth," and is best known for its use in obstetrics to facilitate delivery. Oxytocin-deficient mice deliver their young normally and at about the normal time,[8] but from this we cannot conclude that oxytocin is unimportant in parturition. As John Russell and I wrote in 1998, "The neurohypophysial system regulates the delivery of progeny in all vertebrates, and, during its long evolutionary history, other mechanisms may have evolved convergent roles simply by a process of exclusion. When everything that opposes the actions of oxytocin in parturition is excluded, the things that remain are neutral, assist oxytocin or, in dogging the footsteps of oxytocin, can substitute for it."[9]

Thousands of gene products are involved in the processes by which a few cells in the early embryo proliferate, diversify, migrate, and form the networks that will comprise the mammalian brain. That brain must be robust both to environmental challenges and to intrinsic sources of variation, including the noise in gene expression that comes from the probabilistic nature of gene transcription. Our brains can be robust to even extreme insults in early development, as shown by cases of people with large parts of the brain missing who go on to lead normal lives with little apparent impairment.[10,11] Many mechanisms support that robustness. As mentioned earlier, neurons that receive more or less of a particular chemical signal typically respond by downregulating or upregulating expression of receptors for that chemical, while neurons that receive a stronger or weaker excitatory drive often alter their expression of ion channels to restore a "normal" level of excitability. The distributed nature of information representation means that, even in the developed brains of adults, focal brain injuries tend to degrade performance in a broad domain rather than ablate specific elements of it, and because connections are plastic, considerable functional recovery can occur. Natural selection is likely to favor system architectures that are robust in this way, and systems that have evolved to be robust to extrinsic variation are likely also to be robust to mutations.

Most if not all neurons in the brain produce and secrete one or more peptides as well as conventional neurotransmitters, and express receptors for many peptides. We cannot safely infer that all of these play an important role in brain function. Often, they might be evidence less of the ubiquity of peptide signaling than of the "messiness" of neurons, and of widespread redundancy and degeneracy. I am *not* arguing that what is true in the heart of the brain, where peptides are certainly vitally important, is necessarily true of the whole brain. But in the next and final chapter I will try to pull together the disparate elements of a hypothalamocentric vision of the brain, and contrast it with the classical view.

24 The Tangled Web

A thousand streamers flaunted fair; Various in shape, device, and hue.
—Sir Walter Scott (1771–1832)[1]

The quotation that introduces this chapter is from *Marmion*, an epic poem by Sir Walter Scott about the Battle of Flodden in 1513, and it describes the army raised by King James IV of Scotland from the clans of Scotland. The poem includes the line: "O, what a tangled web we weave..."

I began this book with a reflection that although I, like you, am a human being, a *Homo sapiens*; we are not desiccated calculating machines,[2] but creatures of passion. We share the same genes, with a few minor differences, but those genes did not solely determine who we are now, nor did the differences solely determine the ways in which we differ. The environment into which we were born, our early life experiences and our interactions with close kin, made a big and lasting difference. I asked: how much of our behavior is really governed by reason? How often are the reasons that we give merely self-serving narratives, justifying behaviors governed by things of which we are unaware or only dimly aware or which we prefer not to acknowledge?

I sketched out the "classical" view of the brain—which presents it as a computational structure whose power resides in its scale, in its complexity, in the ability of neurons to compute rapidly, and in the ability of neuronal networks to modify their connections in the light of experience—and asked, where in this edifice are our passions?

Behaviors important to who we are—love and hate, how much we eat and what we eat, how we respond to threat and to stress—are governed by the hypothalamus, but not by the map of how the neurons are connected, rather by where the *receptors* for *peptide signals* are found. The neurons comprise many subpopulations—clans—each with its characteristic phenotype, dictated by the "tartan" of genes that it wears. Clans talk to clans in many different ways with many types of signals on different spatial and temporal scales; they use not one language but many. I argued that these clans have their evolutionary origins in the neurons of "simple" organisms, which combined properties that we have thought of as separate properties of endocrine cells and neurons. They used a diversity of signaling mechanisms, made both peptides and neurotransmitters, and were endowed with a wide range of specialized senses. They had not a single role to which they were committed, but multiple behavioral and physiological functions. As these ancestor neurons proliferated in descendant species, populations differentiated not primarily by gaining properties but by losing properties. Paradoxically, we have become more intelligent as our neurons became less complex. Nevertheless, the neurons of the hypothalamus retain the multifunctionality of their distant ancestors, along with their multitude of sensory abilities.

The classical view of the brain is hierarchical—the "higher centers" integrate, refine, and evaluate information received from more primitive regions. It is a view that ascribes autonomy to neurons, ascribes salience to their electrical activity, and posits that the information carried by each neuron can be understood in isolation from that of all other neurons. Sometimes, this view is argued from evidence that neurons in areas like the cerebral cortex can apparently be extremely selective about the signals to which they respond. But I have argued that populations of neurons in the hypothalamus collectively encode information that is not present in any individual neuron. For example, individual vasopressin neurons fire extremely erratically, and can respond to the sensory signals that are important for vasopressin secretion over only a narrow dynamic range. But because the neurons are heterogeneous, the population as a whole can produce an orderly, sensitive, and proportionate response to physiological challenge; the erratic phasic firing of individual vasopressin cells allows the aggregate population to filter out irrelevant, transient perturbations. In short, the population can do things that the individuals within it cannot.

Peptide signals do not fit the classical view of the brain. Peptides released in the brain play many roles, but three of these I highlighted. First, they can serve as "autoregulators" of neuronal activity—in vasopressin neurons, dynorphin secreted from dendrites is essential for the generation of bursts of activity. Second, peptides can serve as paracrine factors, binding a population together. I gave two examples: oxytocin, released from the dendrites of oxytocin cells by suckling, propagates intense bursts of activity throughout the population of oxytocin cells; vasopressin, released from the dendrites of vasopressin cells, inhibits neighboring vasopressin cells and thereby distributes the burden of a physiological response equitably among them. Third, peptides are so potent and so persistent that intermittent pulsatile secretion from a population can generate a hormonal-like signal within the brain that has effects unconstrained by anatomical connectivity, governed only by the distribution of receptors. Such signals, through effects on gene expression, can alter the phenotype of target neurons, and by priming the release mechanisms of their targets they can produce profound and prolonged changes in functional connectivity.

In considering each of these roles, we can make sense of information transmission in the brain only by considering a *population* of neurons, the clan, as the unit of information processing. Whereas neurotransmitters are whispered secrets that pass from one neuron to another at a very specific time and place, peptides are public announcements, broadcast from one population of neurons to another.

Neurons of the hypothalamus are *not* all alike; individually, they are erratic, messy, quarrelsome, and unreliable. This heterogeneity is not by design but by accident. The patterns of gene expression in any neuron are not rigidly fixed by genetic nature, but are governed by the unique experience of each cell in its life from birth to adulthood. The innervation of each cell is not predetermined with precision. At any one release site, the release of peptide-containing vesicles is a very, very rare event. It is inconceivable that such rare events are rigidly determined by spike activity; there must be a noisy probabilistic relationship between spike activity and these events. It is only the large numbers involved that give an illusion of determinacy—the many endings in a cell, the many cells in the population.

Communication between our brain and the other organs of our bodies goes in both directions. Just as our brains control how our organs behave, so do our organs control how *we* behave, and they do so through the heart of the brain, the hypothalamus. In response to hormonal signals from the adrenal glands and the gonads, some neurons in the hypothalamus don't merely change their electrical activity, they also change the messengers that they produce and who they talk to. The pattern in which they release their messengers not only affects the electrical and secretory activity of their targets, it can also affect the expression of genes in those targets. We are glimpsing a system that is not like a giant supercomputer executing some vast and sophisticated program, but like an ecology of many small analog computers that are constantly reprogramming themselves or being reprogrammed by external events.

Analog computers are now a mere footnote in the history of computing, a history dominated by the rise and rise of digital computers. But when I began as a PhD student in Birmingham, the digital computer available to me was a PDP-9, one of only 445 ever built. It had 32K of core memory, and it filled a small room. At the time I was studying the responses of auditory neurons, and I used that computer to control the frequency and intensity of auditory stimuli and to collect the resulting spike discharge. I could construct a "tuning curve" for an individual neuron that measured the thresholds of response across a range of frequencies in about two minutes—something that otherwise would have taken me about twenty minutes—longer than I could usually maintain a stable recording. But the range of things that could then be done with such primitive machines was limited. When Richard FitzHugh first began studying the Hodgkin-Huxley model of spike generation, to solve the equations on the digital computers then available took a week or more between programming and receiving a solution. So instead he used an analog computer for the work that led to the *FitzHugh-Nagumo model*, a model that became a standard tool in computational neuroscience,[3] and he produced a version of the model that could easily be implemented on the small analog computers then available in student classrooms,[4] including one to which I had access.

Digital computers represent quantities as binary data and process data serially in discrete time steps determined by the clock rate of the central processing unit. To program a digital computer involves writing an algorithm that enables a problem to be solved by a generally very long string of individually simple calculations, successively implemented. By contrast, to program an electronic analog computer involves first representing the problem to be solved by an electrical circuit, and then constructing that circuit—something that the small analog computers made simple by providing elements such as resistors, capacitors, and operational amplifiers to be easily connected in different configurations. In analog computers, quantities are represented as a continuous variable—voltage—and computation is intrinsically parallel, not serial. The results of the computation are instantly available and can be displayed on an oscilloscope in real time. At the time this was an ideal arrangement for studying the theoretical properties of neurons, whose membrane properties are naturally analogous to those of electronic components. Now, the vast increase in the processing power of digital computers has eliminated any advantage that analog computers once had. However, analog computers remain interesting because the way they compute feels much closer to the way the brain works: the brain has no central "clock," all signals exchanged between neurons are continuous, not discrete, and all processing is parallel, not serial.

Reprogramming an analog computer involves physically changing the wiring of its components, an operation that can be seen as analogous to the functional rewiring that occurs in peptide-dependent priming and to the plasticity of gene expression in neurons, as well as to conventional synaptic plasticity. This potential for functional "rewiring"—for reprogramming target populations to fit different physiological states—inspires a different perspective on how the brain works. Computers are programmed once, by an external brain, and the best computers execute that program brilliantly and perfectly. Our brains organize themselves, and reprogram themselves constantly. They guess more than they calculate, they see analogies and jump to conclusions, and they are motivated not by a fixed program but by ephemeral passions and ever-changing needs.

The hormones of the hypothalamus—the heart of the brain—have a particular importance in this process: they are the signals of emotional salience, the links between our passions and our reason, the agents of our urges, our hopes, our fears—of the things that make us human.

Notes

Chapter 1

1. Zeeman, Erik Christopher (1977) *Catastrophe Theory: Selected Papers 1972–1977.* Addison-Wesley.

2. The line "Keats drank confusion to Newton for analyzing the rainbow" is from Skinner, Burrhus Frederic (1971) *Beyond Freedom and Dignity.* Revised edition (2002). Hackett Publishing, 213

3. Keats, John (1820) *Lamia.* Project Gutenberg, http://www.gutenberg.org/ebooks /2490.

4. I don't mean to disparage poets. Robert Burns came to Edinburgh in 1976 to meet the "Blind Poet," Thomas Blacklock, who, in Burns's words "belonged to a set of Critics whose applause I had not even dared to hope." Quoted in Shuttleton DE (2013) "Nae Hottentots": Thomas Blacklock, Robert Burns, and the Scottish vernacular revival. *Eighteenth-Century Life* 37:21–50. Blacklock lived a few hundred yards from where I work.

5. Shakespeare, William. Sonnet 130, "My mistress' eyes are nothing like the sun."

Chapter 2

1. The analogy with clans was introduced in Leng G, Ludwig M (2006) Information processing in the hypothalamus: peptides and analogue computation. *Journal of Neuroendocrinology* 18:379–392.

Chapter 3

1. Andres-Barquin PJ (2002) Santiago Ramón y Cajal and the Spanish school of neurology. *Lancet Neurology* 1:445–452.

2. Tansey EM (1997) Not committing barbarisms: Sherrington and the synapse, 1897. *Brain Research Bulletin* 44:211–212.

3. Hodgkin AL, Huxley AF (1952) A quantitative description of membrane current and its application to conduction and excitation in nerve. *Journal of Physiology* 117:500–544.

4. Berlucchi G, Buchtel HA (2009) Neuronal plasticity: historical roots and evolution of meaning. *Experimental Brain Research* 192:307–319.

5. Llinás RR (1988) The intrinsic electrophysiological properties of mammalian neurons: insights into central nervous system function. *Science* 242:1654–1664.

6. Bean BP (2007) The action potential in mammalian central neurons. *Nature Reviews Neuroscience* 8:451–465.

7. Lettvin JY, Maturana HR, McCulloch WS, Pitts WH (1959) What the frog's eye tells the frog's brain. *Proceedings of the Institute of Radio Engineers* 47:1940–1951.

8. Gasser HS (1937) The control of excitation in the nervous system. *Bulletin of the New York Academy of Medicine* 13:324–348.

9. Moore D (1965) Cramming more components onto integrated circuits. *Electronics Magazine*, 19 April.

10. McCulloch W, Pitts W (1943) A logical calculus of ideas immanent in nervous activity. *Bulletin of Mathematical Biology* 5:115–133.

Chapter 4

1. Hume, David (2000) *A Treatise of Human Nature*. Edited by David Fate Norton and Mary J. Norton. Oxford University Press, 266.

2. Hobbes, Thomas (1651). *De Cive: The English Version*, entitled, in the first edition, *Philosophicall Rudiments Concerning Government and Society*. Blackmask Online (2000) P5. http://www.unilibrary.com/ebooks/Hobbes,%20Thomas%20-%20De%20Cive.pdf. Accessed 14/11/2017.

3. Popper, Karl (1963) *Conjectures and Refutations: The Growth of Scientific Knowledge*. Routledge, 69.

4. Popper, Karl (2002) *The Logic of Scientific Discovery*. Routledge Classics, 93–94.

5. Stevenson, Robert Louis (1848) *Edinburgh: Picturesque Notes*. Transcribed by David Price from the 1903 Seeley & Co. Ltd. edition. Project Gutenberg, http://www.gutenberg.org/ebooks/382.

6. Hume, David (1748/1902) *An Enquiry Concerning Human Understanding*. Second edition. Project Gutenberg, http://www.gutenberg.org/ebooks/9662.

7. Hume, David *An Enquiry*.... sec. IV, para 24

8. Hume, David *An Enquiry*... sec. IV, para. 26.

8. Popper, Karl (2002) *The Logic of Scientific Discovery*. Routledge Classics, 280.

9. Hume, David (1748) *An Enquiry*, sec. V, para. 45.

Chapter 5

1. Harris, Geoffrey W (1955) *Neural Control of the Pituitary Gland*. Monographs of the Physiological Society. Edward Arnold.

2. Leng G, Pineda R, Sabatier N, Ludwig M (2015) 60 years of neuroendocrinology: The posterior pituitary, from Geoffrey Harris to our present understanding. *Journal of Endocrinology* 226:T173–185.

3. Russell JA, Leng G (1998) Sex, parturition and motherhood without oxytocin? *Journal of Endocrinology* 157:343–359.

4. Cross BA, Harris GW (1950) Milk ejection following electrical stimulation of the pituitary stalk in rabbits. *Nature* 166:994–995.

5. Cross BA, Harris GW (1952) The role of the neurohypophysis in the milk-ejection reflex. *Journal of Endocrinology* 8:148–161.

6. Yagi K, Azuma T, Matsuda K (1966) Neurosecretory cell: capable of conducting impulse in rats. *Science* 154:778–779.

7. Wakerley JB, Lincoln DW (1973) The milk-ejection reflex of the rat: a 20- to 40-fold acceleration in the firing of paraventricular neurones during oxytocin release. *Journal of Endocrinology* 57:477–493.

8. Lincoln DW, Wakerley JB (1974) Electrophysiological evidence for the activation of supraoptic neurones during the release of oxytocin. *Journal of Physiology* 242:533–554.

9. Harris GW, Manabe Y, Ruf KB (1969) A study of the parameters of electrical stimulation of unmyelinated fibers in the pituitary stalk. *Journal of Physiology* 203:67–81.

10. Bicknell RJ (1988) Optimizing release from peptide hormone secretory nerve terminals. *Journal of Experimental Biology* 139:51–65.

11. Ferguson JKW (1941) A study of the motility of the intact uterus at term. *Surgery, Gynecology and Obstetrics* 73:359–366.

12. Dey FL, Fisher C, Ransom SW (1941) Disturbance in pregnancy and labor in guinea-pigs with hypothalamic lesions. *American Journal of Obstetrics and Gynecology* 42:459–466.

13. Gunther M (1948) The posterior pituitary and labor. *British Medical Journal* 1:567.

14. Melander SE (1961) Oxytocinase activity of plasma of pregnant women. *Nature* 191:176–177.

15. O'Byrne KT, Ring JP, Summerlee AJ (1986) Plasma oxytocin and oxytocin neurone activity during delivery in rabbits. *Journal of Physiology* 370:501–513.

16. Paisley AC, Summerlee AJ (1984) Activity of putative oxytocin neurones during reflex milk ejection in conscious rabbits. *Journal of Physiology* 347:465–478.

17. Summerlee AJ, Lincoln DW (1981) Electrophysiological recordings from oxytocinergic neurones during suckling in the unanaesthetized lactating rat. *Journal of Endocrinology* 90:255–265.

18. Summerlee AJ (1981) Extracellular recordings from oxytocin neurones during the expulsive phase of birth in unanaesthetized rats. *Journal of Physiology* 321:1–9.

19. Fuchs AR, Romero R, Keefe D, Parra M, Oyarzun E, Behnke E (1991) Oxytocin secretion and human parturition: pulse frequency and duration increase during spontaneous labor in women. *American Journal of Obstetrics and Gynecology* 165:1515–1523.

20. Luckman SM, Antonijevic I, Leng G, Dye S, Douglas AJ, Russell JA, Bicknell RJ (1993) The maintenance of normal parturition in the rat requires neurohypophysial oxytocin. *Journal of Neuroendocrinology* 5:7–12.

21. Parry LJ, Bathgate RA (2000) The role of oxytocin and regulation of uterine oxytocin receptors in pregnant marsupials. *Experimental Physiology* 85 S:91S–99S.

22. Moss RL, Dyball RE, Cross BA (1972) Excitation of antidromically identified neurosecretory cells of the paraventricular nucleus by oxytocin applied iontophoretically. *Experimental Neurology* 34:95–102.

23. Mens WB, Witter A, van Wimersma Greidanus TB (1983) Penetration of neurohypophyseal hormones from plasma into cerebrospinal fluid (CSF): half-times of disappearance of these neuropeptides from CSF. *Brain Research* 262:143–149.

24. Freund-Mercier MJ, Richard P (1981) Excitatory effects of intraventricular injections of oxytocin on the milk ejection reflex in the rat. *Neuroscience Letters* 23:193–198.

25. Richard P, Moos F, Dayanithi G, Gouzènes L, Sabatier N (1997) Rhythmic activities of hypothalamic magnocellular neurons: autocontrol mechanisms. *Biology of the Cell* 89:555–560.

Chapter 6

1. Yalow RS (1978) Radioimmunoassay: a probe for the fine structure of biologic systems. *Science* 200:1236–1245.

2. Dierschke DJ, Bhattacharya AN, Atkinson LE, Knobil E (1970) Circhoral oscillations of plasma LH levels in the ovariectomized rhesus monkey. *Endocrinology* 87:850–853.

3. Santen RJ, Bardin CW (1973) Episodic luteinizing hormone secretion in man: pulse analysis, clinical interpretation, physiologic mechanisms. *Journal of Clinical Investigation* 52:2617–2628.

4. Guillemin R (1978) Peptides in the brain: the new endocrinology of the neuron. *Science* 202:390–402.

5. Schally AV (1978) Aspects of hypothalamic regulation of the pituitary gland. *Science* 202:18–28.

6. Morgan K, Millar RP (2004) Evolution of GnRH ligand precursors and GnRH receptors in protochordate and vertebrate species. *General and Comparative Endocrinology* 139:191–197.

7. Clarke IJ, Cummins JT (1982) The temporal relationship between gonadotropin releasing hormone (GnRH) and luteinizing hormone (LH) secretion in ovariectomized ewes. *Endocrinology* 111:1737–1739.

8. Moenter SM, Brand RC, Karsch FJ (1992) Dynamics of gonadotropin-releasing hormone (GnRH) secretion during the GnRH surge: insights into the mechanism of GnRH surge induction. *Endocrinology* 130:2978–2984.

9. Knobil E (2005) Discovery of the hypothalamic gonadotropin-releasing hormone pulse generator and of its physiologic significance. *American Journal of Obstetrics and Gynecology* 193:1765–1766.

10. Lucas X (2014) Clinical use of deslorelin (GnRH agonist) in companion animals: a review. *Reproduction in Domestic Animals* 49 Suppl 4:64–71.

11. Hayes FJ, Crowley WF Jr (1998) Gonadotropin pulsations across development. *Hormone Research* 49:163–168.

12. Heape W (1900) The "sexual season" of mammals and the relation of the "prooestrum" to menstruation. *Quarterly Journal of Microscopic Science* 44:1–70.

13. Fink G (2015) 60 years of neuroendocrinology. Memoir: Harris' neuroendocrine revolution: of portal vessels and self-priming. *Journal of Endocrinology* 226:T13–T24.

14. Lewis CE, Morris JF, Fink G (1985) The role of microfilaments in the priming effect of LH-releasing hormone: an ultrastructural study using cytochalasin B. *Journal of Endocrinology* 106:211–218.

15. Scullion S, Brown D, Leng G (2004) Modeling the pituitary response to luteinizing hormone-releasing hormone. *Journal of Neuroendocrinology* 16:265–271.

16. Robinson IC (1991) The growth hormone secretory pattern: a response to neuroendocrine signals. *Acta Paediatrica Scandinavica Supplement* 372:70–78.

17. Walker JJ, Spiga F, Waite E, Zhao Z, Kershaw Y, Terry JR, Lightman SL (2012) The origin of glucocorticoid hormone oscillations. *PLoS Biology* 10(6):e1001341.

Chapter 7

1. Poulain DA, Wakerley JB (1986) Afferent projections from the mammary glands to the spinal cord in the lactating rat—II. Electrophysiological responses of spinal neurons during stimulation of the nipples, including suckling. *Neuroscience* 19:511–521.

2. Fénelon VS, Poulain DA (1992) Electrical activity of dorsal horn neurons during the suckling-induced milk ejection reflex in the lactating rat. *Journal of Neuroendocrinology* 4:575–584.

3. Sutherland RC, Juss TS, Wakerley JB (1987) Prolonged electrical stimulation of the nipples evokes intermittent milk ejection in the anaesthetised lactating rat. *Experimental Brain Research* 66:29–34.

4. Hoffman GE, Smith MS, Verbalis JG (1993) c-Fos and related immediate early gene products as markers of activity in neuroendocrine systems. *Frontiers in Neuroendocrinology* 14:173–213.

5. Moos F, Marganiec A, Fontanaud P, Guillou-Duvoid A, Alonso G (2004) Synchronization of oxytocin neurons in suckled rats: possible role of bilateral innervation of hypothalamic supraoptic nuclei by single medullary neurons. *European Journal of Neuroscience* 20:66–78.

6. Lambert RC, Moos FC, Richard P (1993) Action of endogenous oxytocin within the paraventricular or supraoptic nuclei: a powerful link in the regulation of the bursting pattern of oxytocin neurons during the milk-ejection reflex in rats. *Neuroscience* 57:1027–1038.

7. Freund-Mercier MJ, Moos F, Poulain DA, Richard P, Rodriguez F, Theodosis DT, Vincent JD (1988) Role of central oxytocin in the control of the milk ejection reflex. *Brain Research Bulletin* 20:737–741.

8. Freund-Mercier MJ, Stoeckel ME, Palacios JM, Pazos A, Reichhart JM, Porte A, Richard P (1987) Pharmacological characteristics and anatomical distribution of

[3H]oxytocin-binding sites in the Wistar rat brain studied by autoradiography. *Neuroscience* 20:599–614.

9. Freund-Mercier MJ, Stoeckel ME, Klein MJ (1994) Oxytocin receptors on oxytocin neurones: histoautoradiographic detection in the lactating rat. *Journal of Physiology* 480:155–161.

10. Theodosis DT (1985) Oxytocin-immunoreactive terminals synapse on oxytocin neurones in the supraoptic nucleus. *Nature* 313:682–684.

11. Dyball RE, Leng G (1986) Regulation of the milk ejection reflex in the rat. *Journal of Physiology* 380:239–256.

12. Leng G, Dyball RE (1984) Recurrent inhibition: a recurring misinterpretation. *Quarterly Journal of Experimental Physiology* 69:393–395.

13. Andrew RD (1986) Intrinsic membrane properties of magnocellular neurosecretory neurons recorded in vitro. *Federation Proceedings* 45:2306–2311.

14. Leng G (1981) The effects of neural stalk stimulation upon firing patterns in rat supraoptic neurones. *Experimental Brain Research* 41:135–45.

15. Morris JF, Pow DV (1988) Capturing and quantifying the exocytotic event. *Journal of Experimental Biology* 139:81–103.

16. Ludwig M, Sabatier N, Bull PM, Landgraf R, Dayanithi G, Leng G (2002) Intracellular calcium stores regulate activity-dependent neuropeptide release from dendrites. *Nature* 418:85–89.

17. Theodosis DT (2002) Oxytocin-secreting neurons: a physiological model of morphological neuronal and glial plasticity in the adult hypothalamus. *Frontiers in Neuroendocrinology* 23:101–135.

18. Hatton GI, Tweedle CD (1982) Magnocellular neuropeptidergic neurons in hypothalamus: increases in membrane apposition and number of specialized synapses from pregnancy to lactation. *Brain Research Bulletin* 8:197–204.

19. Theodosis DT (2002) Oxytocin-secreting neurons: a physiological model of morphological neuronal and glial plasticity in the adult hypothalamus. *Frontiers in Neuroendocrinology* 23:101–135.

20. Leng G, Moos FC, Armstrong WE (2010) The adaptive brain: Glenn Hatton and the supraoptic nucleus. *Journal of Neuroendocrinology* 22:318–329.

21. Catheline G, Touquet B, Lombard MC, Poulain DA, Theodosis DT (2006) A study of the role of neuro-glial remodeling in the oxytocin system at lactation. *Neuroscience* 137:309–316.

22. Hatton GI, Wang YF (2008) Neural mechanisms underlying the milk ejection burst and reflex. *Progress in Brain Research* 170:155–166.

23. Israel JM, Oliet SH, Ciofi P (2016) Electrophysiology of hypothalamic magnocellular neurons in vitro: a rhythmic drive in organotypic cultures and acute slices. *Frontiers in Neuroscience* 10:109.

24. Chevaleyre V, Moos FC, Desarménien MG (2001) Correlation between electrophysiological and morphological characteristics during maturation of rat supraoptic neurons. *European Journal of Neuroscience* 13:1136–1146.

25. Negoro H, Uchide K, Honda K, Higuchi T (1985) Facilitatory effect of antidromic stimulation on milk ejection-related activation of oxytocin neurons during suckling in the rat. *Neuroscience Letters* 59:21–25.

26. Belin V, Moos F (1986) Paired recordings from supraoptic and paraventricular oxytocin cells in suckled rats: recruitment and synchronization. *Journal of Physiology* 377:369–390.

27. Moos FC (1995) GABA-induced facilitation of the periodic bursting activity of oxytocin neurones in suckled rats. *Journal of Physiology* 488:103–114.

28. Leng G, Caquineau C, Ludwig M (2008) Priming in oxytocin cells and in gonadotrophs. *Neurochemistry Research* 33:668–677.

29. Ludwig M, Sabatier N, Bull PM, Landgraf R, Dayanithi G, Leng G (2002) Intracellular calcium stores regulate activity-dependent neuropeptide release from dendrites. *Nature* 418:85–89.

30. Tobin VA, Hurst G, Norrie L, Dal Rio FP, Bull PM, Ludwig M (2004) Thapsigargin-induced mobilization of dendritic dense-cored vesicles in rat supraoptic neurons. *European Journal of Neuroscience* 19:2909–2912.

31. Tobin VA, Ludwig M (2007) The role of the actin cytoskeleton in oxytocin and vasopressin release from rat supraoptic nucleus neurons. *Journal of Physiology* 582:1337–1348.

32. Sabatier N (2006) Alpha-melanocyte-stimulating hormone and oxytocin: a peptide signaling cascade in the hypothalamus. *Journal of Neuroendocrinology* 18:703–710.

33. Rossoni E, Feng J, Tirozzi B, Brown D, Leng G, Moos F (2008) Emergent synchronous bursting of oxytocin neuronal network. *PLoS Computational Biology* 4(7):e1000123.

34. Honda K, Sudo A, Ikeda K (2013) Oxytocin cells in the supraoptic nucleus receive excitatory synaptic inputs from the contralateral supraoptic and paraventricular nuclei in the lactating rat. *Journal of Reproduction and Development* 59:569–574.

Chapter 8

1. Clough, Arthur Hugh (1888) *Poems of Arthur Hugh Clough*. Macmillan.

2. Casoni F, Malone SA, Belle M, Luzzati F, Collier F, Allet C, Hrabovszky E et al. (2016) Development of the neurons controlling fertility in humans: new insights from 3D imaging and transparent fetal brains. *Development* 143:3969–3981.

3. Forni PE, Wray S (2015) GnRH, anosmia and hypogonadotropic hypogonadism: where are we? *Frontiers in Neuroendocrinology* 36:165–177.

4. Kallmann F, Schoenfeld WA, Barrera SE (1944) The genetic aspects of primary eunuchoidism. *American Journal of Mental Deficiency* 48:203–236.

5. Halász B, Pupp L (1965) Hormone secretion of the anterior pituitary gland after physical interruption of all nervous pathways to the hypophysiotrophic area. *Endocrinology* 77:553–562.

6. Knobil E (1989) The electrophysiology of the GnRH pulse generator in the rhesus monkey. *Journal of Steroid Biochemistry* 33:669–671.

7. O'Byrne K, Knobil E (1993) Electrophysiological approaches to gonadotrophin releasing hormone pulse generator activity in the rhesus monkey. *Human Reproduction* 8 Suppl 2:37–40.

8. Tanaka T, Mori Y, Hoshino K (1992) Hypothalamic GnRH pulse generator activity during the estradiol-induced LH surge in ovariectomized goats. *Neuroendocrinology* 56:641–645.

9. Nishihara M, Sano A, Kimura F (1994) Cessation of the electrical activity of gonadotropin-releasing hormone pulse generator during the steroid-induced surge of luteinizing hormone in the rat. *Neuroendocrinology* 59:513–519.

10. Goubillon ML, Strutton PH, O'Byrne KT, Thalabard JC, Coen CW (1999) Ketamine-induced general anesthesia is compatible with gonadotropin-releasing hormone pulse generator activity in gonadectomized rats: prospects for detailed electrophysiological studies in vivo. *Brain Research* 841:197–201.

11. Rinzel J, Lee YS (1987) Dissection of a model for neuronal parabolic bursting. *Journal of Mathematical Biology* 25:653–675.

12. Canavier CC, Clark JW, Byrne JH (1991) Simulation of the bursting activity of neuron R15 in Aplysia: role of ionic currents, calcium balance, and modulatory transmitters. *Journal of Neurophysiology* 66:2107–2124.

13. Weiner RI, Wetsel W, Goldsmith P, Martinez de la Escalera G, Windle J, Padula C, Choi A, Negro-Vilar A, Mellon P (1992) Gonadotropin-releasing hormone neuronal cell lines. *Frontiers in Neuroendocrinology* 13:95–119.

14. Charles AC, Hales TG (1995) Mechanisms of spontaneous calcium oscillations and action potentials in immortalized hypothalamic (GT1–7) neurons. *Journal of Neurophysiology* 73:56–64.

15. Nunemaker CS, DeFazio RA, Geusz ME, Herzog ED, Pitts GR, Moenter SM (2001) Long-term recordings of networks of immortalized GnRH neurons reveal episodic patterns of electrical activity. *Journal of Neurophysiology* 86:86–93.

16. Terasawa E, Schanhofer WK, Keen KL, Luchansky L (1999) Intracellular Ca^{2}+ oscillations in luteinizing hormone-releasing hormone neurons derived from the embryonic olfactory placode of the rhesus monkey. *Journal of Neuroscience* 19:5898–5909.

17. Terasawa E, Keen KL, Mogi K, Claude P (1999) Pulsatile release of luteinizing hormone-releasing hormone (LHRH) in cultured LHRH neurons derived from the embryonic olfactory placode of the rhesus monkey. *Endocrinology* 140:1432–1441.

18. Constantin S, Caraty A, Wray S, Duittoz AH (2009) Development of gonadotropin-releasing hormone-1 secretion in mouse nasal explants. *Endocrinology* 150:3221–3227.

19. Abe H, Terasawa E (2005) Firing pattern and rapid modulation of activity by estrogen in primate luteinizing hormone releasing hormone-1 neurons. *Endocrinology* 146:4312–4320.

20. Tsien RW, Tsien RY (1990) Calcium channels, stores, and oscillations. *Annual Reviews of Cell Biology* 6:715–760.

21. Constantin S, Klenke U, Wray S (2010) The calcium oscillator of GnRH-1 neurons is developmentally regulated. *Endocrinology* 151: 3863–3873.

22. Campbell RE, Han SK, Herbison AE (2005) Biocytin filling of adult gonadotropin-releasing hormone neurons in situ reveals extensive, spiny, dendritic processes. *Endocrinology* 146:1163–1169.

23. Campbell RE, Gaidamaka G, Han SK, Herbison AE (2009) Dendro-dendritic bundling and shared synapses between gonadotropin-releasing hormone neurons. *Proceedings of the National Academy of Sciences USA* 106:10835–10840.

24. Theodosis DT, Chapman DB, Montagnese C, Poulain DA, Morris JF (1986) Structural plasticity in the hypothalamic supraoptic nucleus at lactation affects oxytocin-, but not vasopressin-secreting neurones. *Neuroscience* 17:661–678.

25. Han SK, Lee K, Bhattarai JP, Herbison AE (2010) Gonadotrophin-releasing hormone (GnRH) exerts stimulatory effects on GnRH neurons in intact adult male and female mice. *Journal of Neuroendocrinology* 22:188–195.

26. Constantin S, Iremonger KJ, Herbison AE (2013) In vivo recordings of GnRH neuron firing reveal heterogeneity and dependence upon $GABA_A$ receptor signaling. *Journal of Neuroscience* 33:9394–9401.

27. Duittoz A, Cognié J, Decourt C, Derouin F, Forestier A, Lecomte F, Bouakkaz A, Reigner F (2018) The horse: an unexpected animal model for (unexpected) neuroendocrinology. In Ludwig M and Levkowitz G (eds) *Masterclasses in Neuroendocrinology*. Wiley. In press.

Chapter 9

1. McNeilly AS, Crawford JL, Taragnat C, Nicol L, McNeilly JR (2003) The differential secretion of FSH and LH: regulation through genes, feedback and packaging. *Reproduction Supplement* 61:463–476.

2. Hrabovszky E, Shughrue PJ, Merchenthaler I, Hajszan T, Carpenter CD, Liposits Z, Petersen SL (2000) Detection of estrogen receptor-β messenger ribonucleic acid and ^{125}I-estrogen binding sites in luteinizing hormone-releasing hormone neurons of the rat brain. *Endocrinology* 141:3506–3509.

3. Maeda K, Ohkura S, Uenoyama Y, Wakabayashi Y, Oka Y, Tsukamura H, Okamura H (2010) Neurobiological mechanisms underlying GnRH pulse generation by the hypothalamus. *Brain Research* 1364:103–115.

4. Smith JT (2013) Sex steroid regulation of kisspeptin circuits. *Advances in Experimental Medicine and Biology* 784:275–295.

5. Steiner RA (2013) Kisspeptin: past, present, and prologue. *Advances in Experimental Medicine and Biology* 784:3–7.

6. Herbison AE (2016) Control of puberty onset and fertility by gonadotropin-releasing hormone neurons. *Nature Reviews Endocrinology* 12:452–466.

7. Plant TM (2015) Neuroendocrine control of the onset of puberty. *Frontiers in Neuroendocrinology* 38:73–88.

8. Topaloglu AK, Kotan LD (2016) Genetics of hypogonadotropic hypogonadism. *Endocrine Development* 29:36–49.

9. Ramaswamy S, Guerriero KA, Gibbs RB, Plant TM (2008) Structural interactions between kisspeptin and GnRH neurons in the mediobasal hypothalamus of the male rhesus monkey (*Macaca mulatta*) as revealed by double immunofluorescence and confocal microscopy. *Endocrinology* 149:4387–4395.

10. Keen KL, Wegner FH, Bloom SR, Ghatei MA, Terasawa E (2008) An increase in kisspeptin-54 release occurs with the pubertal increase in luteinizing hormone-releasing hormone-1 release in the stalk-median eminence of female rhesus monkeys in vivo. *Endocrinology* 149:4151–4157.

11. Boyden ES, Zhang F, Bamberg E, Georg Nagel G, Deisseroth K (2005) Millisecond-timescale, genetically targeted optical control of neural activity. *Nature Neuroscience* 8:1263–1268.

12. Han SY, McLennan T, Czieselsky K, Herbison AE (2015) Selective optogenetic activation of arcuate kisspeptin neurons generates pulsatile luteinizing hormone secretion. *Proceedings of the National Academy of Sciences USA* 112:13109–13114.

13. Li X-F, Kinsey-Jones JS, Cheng Y, Knox AM, Lin Y, Petrou NA, Roseweir A, Lightman SL, Milligan SR, Millar RP, O'Byrne KT (2009) Kisspeptin signaling in the hypothalamic arcuate nucleus regulates GnRH pulse generator frequency in the rat. *PLoS One* 4(12): e8334.

14. Iremonger KJ, Herbison AE (2015) Multitasking in gonadotropin-releasing hormone neuron dendrites. *Neuroendocrinology* 102:1–7.

15. Iremonger KJ, Porteous R, Herbison AE (2017) Spike and neuropeptide-dependent mechanisms control GnRH neuron nerve terminal Ca^2+ over diverse time scales. *Journal of Neuroscience* 37:3342–3351.

16. Moenter SM, Caraty A, Locatelli A, Karsch FJ (1991) Pattern of gonadotropin-releasing hormone (GnRH) secretion leading up to ovulation in the ewe: existence of a preovulatory GnRH surge. *Endocrinology* 129:1175–1182.

17. Kelly MJ, Zhang C, Qiu J, Rønnekleiv OK (2013) Pacemaking kisspeptin neurons. *Experimental Physiology* 98:1535–1543.

18. Herbison AE, Porteous R, Pape JR, Mora JM, Hurst PR (2008) Gonadotropin-releasing hormone neuron requirements for puberty, ovulation, and fertility. *Endocrinology* 149:597–604.

19. Zhang G, Li J, Purkayastha S, Tang Y, Zhang H, Yin Y, Li B, Liu G, Cai D (2013) Hypothalamic programming of systemic aging involving IKK-β, NF-κB and GnRH. *Nature* 497:211–216.

Chapter 10

1. Mason WT (1983) Electrical properties of neurons recorded from the rat supraoptic nucleus in vitro. *Proceedings of the Royal Society of London B* 217:141–161.

2. MacGregor DJ, Leng G (2013) Spike triggered hormone secretion in vasopressin cells: a model investigation of mechanism and heterogeneous population function. *PLoS Computational Biology* 9(8):e1003187.

3. Dunn FL, Brennan TJ, Nelson AE, Robertson GL (1973) The role of blood osmolality and volume in regulating vasopressin secretion in the rat. *Journal of Clinical Investigation* 52: 3212–3219.

4. Leng G, Dyball REJ (1987) The role of supraoptic neurones in blood pressure regulation. In Ciriello J, Calaresu FR, Renaud LP, Polosa C (eds) *Organization of the Autonomic Nervous System: Central and Peripheral Mechanisms*. Alan R. Liss, 447–456.

5. Madore BF, Freedman WL (1983) Computer simulations of the Belousov-Zhabotinsky reaction. *Science* 222:615–616.

6. Leng G, Brown D (1997) The origins and significance of pulsatility in hormone secretion from the pituitary. *Journal of Neuroendocrinology* 9:493–513.

7. Armstrong WE, Stern JE (1998) Phenotypic and state-dependent expression of the electrical and morphological properties of oxytocin and vasopressin neurones. *Progress in Brain Research* 119:101–113.

8. Brown CH, Bourque CW (2006) Mechanisms of rhythmogenesis: insights from hypothalamic vasopressin neurons. *Trends in Neuroscience* 29:108–115.

9. Macgregor DJ, Leng G. (2012) Phasic firing in vasopressin cells: understanding its functional significance through computational models *PLoS Computational Biology* 8(10):e1002740.

10. Clayton TF, Murray AF, Leng G (2010) Modeling the in vivo spike activity of phasically-firing vasopressin cells. *Journal of Neuroendocrinology* 22:1290–1300.

11. MacGregor DJ, Leng G. (2013) Information coding in vasopressin neurons: the role of asynchronous bistable burst firing. *Biosystems* 112:85–93.

12. Sabatier N, Leng G. (2007) Bistability with hysteresis in the activity of vasopressin cells. *Journal of Neuroendocrinology* 19:95–101.

13. Leng G, Brown C, Sabatier N, Scott V (2008) Population dynamics in vasopressin neurons. *Neuroendocrinology* 88:160–172.

14. Mason WT (1980) Supraoptic neurones of rat hypothalamus are osmosensitive. *Nature* 287:154–157.

15. Bourque CW (2008) Central mechanisms of osmosensation and systemic osmoregulation. *Nature Reviews Neuroscience* 9:519–531.

16. Douglass JK, Wilkens L, Pantazelou E, Moss F (1993) Noise enhancement of information transfer in crayfish mechanoreceptors by stochastic resonance. *Nature* 365:337–340.

17. McDonnell MD, Ward LM (2011) The benefits of noise in neural systems: bridging theory and experiment. *Nature Reviews Neuroscience* 12:415–426.

18. Leng G, Mason WT, Dyer RG (1982) The supraoptic nucleus as an osmoreceptor. *Neuroendocrinology* 34:75–82.

19. Leng G, Blackburn RE, Dyball RE, Russell JA (1989) Role of anterior peri-third ventricular structures in the regulation of supraoptic neuronal activity and neurohypophysial hormone secretion in the rat. *Journal of Neuroendocrinology* 1:35–46.

20. McCormick DA (2005) Neuronal networks: flip-flops in the brain. *Current Biology* 15:R294–296.

Chapter 11

1. Verney EB (1946) Absorption and excretion of water; the antidiuretic hormone. *Lancet* 2: 781–783.

2. Agre P (2006) The aquaporin water channels. *Proceedings of the American Thoracic Society* 3:5–13.

3. Valtin H (1982) The discovery of the Brattleboro rat, recommended nomenclature, and the question of proper controls. *Annals of the New York Academy of Sciences* 394:1–9.

4. Arima H, Oiso Y (2010) Mechanisms underlying progressive polyuria in familial neurohypophysial diabetes insipidus. *Journal of Neuroendocrinology* 22:754–757.

5. Cowen LE, Hodak SP, Verbalis JG (2013) Age-associated abnormalities of water homeostasis. *Endocrinology and Metabolism Clinics of North America* 42:349–370.

6. Leng G, Sabatier N (2016) Measuring oxytocin and vasopressin: bioassays, immunoassays and random numbers. *Journal of Neuroendocrinology* 28(10). doi: 10.1111/jne.12413.

7. Leng G, Dyball RE, Luckman SM (1992) Mechanisms of vasopressin secretion. *Hormone Research* 37:33–38.

8. Bicknell RJ (1988) Optimizing release from peptide hormone secretory nerve terminals. *Journal of Experimental Biology* 139:51–65.

9. MacGregor DJ, Leng G (2013) Spike triggered hormone secretion in vasopressin cells: a model investigation of mechanism and heterogeneous population function. *PLoS Computational Biology* 9(8):e1003187.

10. Brown CH, Ruan M, Scott V, Tobin VA, Ludwig M (2008) Multi-factorial somatodendritic regulation of phasic spike discharge in vasopressin neurons. *Progress in Brain Research* 170:219–228.

11. Shadlen MN, Newsome WT (1994) Noise, neural codes and cortical organization. *Current Opinion in Neurobiology* 4:569–579.

Chapter 12

1. Gould, Stephen Jay (1990) *An Urchin in the Storm*. Penguin Books, 97.

2. Leng G, MacGregor DJ (2008) Mathematical modeling in neuroendocrinology. *Journal of Neuroendocrinology* 20:713–718.

3. Morris JF (1976) Distribution of neurosecretory granules among the anatomical compartments of the neurosecretory processes of the pituitary gland: a quantitative

ultrastructural approach to hormone storage in the neural lobe. *Journal of Endocrinology* 68:225–234.

4. Avogadro's number, 6.02×10^{23}, is the number of molecules of a specific substance present in 1 mole of that substance (1 mole = W grams for a molecule with molecular weight W).

5. Nordmann JJ, Morris JF (1984) Method for quantitating the molecular content of a subcellular organelle: hormone and neurophysin content of newly formed and aged neurosecretory granules. *Proceedings of the National Academy of Sciences USA* 81:180–184.

6. Morris JF (1976) Hormone storage in individual neurosecretory granules of the pituitary gland: a quantitative ultrastructural approach to hormone storage in the neural lobe. *Journal of Endocrinology* 68:209–224.

7. Nordmann JJ (1977) Ultrastructural morphometry of the rat neurohypophysis. *Journal of Anatomy* 123:213–218.

8. Leng G, Ludwig M (2008) Neurotransmitters and peptides: whispered secrets and public announcements. *Journal of Physiology* 586:5625–5632.

9. Leng G, Sabatier N (2016) Measuring oxytocin and vasopressin: bioassays, immunoassays and random numbers. *Journal of Neuroendocrinology* 28(10). doi: 10.1111/jne.12413.

10. Nordmann JJ (1985). Hormone content and movement of neurosecretory granules in the rat neural lobe during and after dehydration. *Neuroendocrinology* 40:25–32.

Chapter 13

1. Lord John Russell, in a letter to Thomas Attwood, October 1831, after the rejection in the House of Lords of the Reform Bill.

2. This chapter is based on Leng G, Ludwig M (2008) Neurotransmitters and peptides: whispered secrets and public announcements. *Journal of Physiology* 586:5625–32.

3. Kimura T, Makino Y, Bathgate R, Ivell R, Nobunaga T, Kubota Y, Kumazawa I et al. (1997) The role of N-terminal glycosylation in the human oxytocin receptor. *Molecular Human Reproduction* 3:957–963.

4. Abbott NJ (2004) Evidence for bulk flow of brain interstitial fluid: significance for physiology and pathology. *Neurochemistry International* 45:545–552.

Chapter 14

1. Rosch PJ (1999) Reminiscences of Hans Selye, and the birth of "stress." *International Journal of Emergency Mental Health* 1:59–66.

2. Spiga F, Walker JJ, Gupta R, Terry JR, Lightman SL (2015) 60 years of neuroendocrinology. Glucocorticoid dynamics: insights from mathematical, experimental and clinical studies. *Journal of Endocrinology* 226:T55–66.

3. Gillies G, Lowry P (1979) Corticotrophin releasing factor may be modulated vasopressin. *Nature* 278:463–464.

4. Vale W, Spiess J, Rivier C, Rivier J (1981) Characterization of a 41-residue ovine hypothalamic peptide that stimulates secretion of corticotropin and beta-endorphin. *Science* 213:1394–1397.

5. Gillies GE, Linton EA, Lowry PJ (1982) Corticotropin releasing activity of the new CRF is potentiated several times by vasopressin. *Nature* 299:355–357.

6. Sawchenko PE, Swanson LW, Vale WW (1984) Co-expression of corticotropin-releasing factor and vasopressin immunoreactivity in parvocellular neurosecretory neurons of the adrenalectomized rat. *Proceedings of the National Academy of Sciences USA* 81:1883–1887.

7. Ma XM, Lightman SL (1978) The arginine vasopressin and corticotrophin-releasing hormone gene transcription responses to varied frequencies of repeated stress in rats. *Journal of Physiology* 510:605–614.

8. Featherstone K, White MR, Davis JR (2012) The prolactin gene: a paradigm of tissue-specific gene regulation with complex temporal transcription dynamics. *Journal of Neuroendocrinology* 24:977–990.

9. Featherstone K, Hey K, Momiji H, McNamara AV, Patist AL, Woodburn J, Spiller DG et al. (2016) Spatially coordinated dynamic gene transcription in living pituitary tissue. *eLife* 5:e08494.

10. Eberhart JA, Yodyingyuad U, Keverne EB (1985) Subordination in male talapoin monkeys lowers sexual behavior in the absence of dominants. *Physiology and Behavior* 35:673–677.

11. Grattan DR (2015) 60 years of neuroendocrinology. The hypothalamo-prolactin axis. *Journal of Endocrinology* 226:T101–122.

Chapter 15

1. The organization of the suprachiasmatic nucleus is covered well in Evans JA (2016) Collective timekeeping among cells of the master circadian clock, *Journal of Endocrinology* 230:R27–49, and Antle MC, Silver R (2005) Orchestrating time: arrangements of the brain circadian clock, *Trends in Neuroscience* 28:145–151.

2. Ralph MR, Menaker M (1988) A mutation of the circadian system in golden hamsters. *Science* 241:1225–1227.

3. Silver R, LeSauter J, Tresco PA, Lehman MN (1996) A diffusible coupling signal from the transplanted suprachiasmatic nucleus controlling circadian locomotor rhythms. *Nature* 382:810–813.

4. Tei H, Okamura H, Shigeyoshi Y, Fukuhara C, Ozawa R, Hirose M, Sakaki Y (1997) Circadian oscillation of a mammalian homolog of the *Drosophila period* gene. *Nature* 389:512–516.

5. Herzog ED, Tosini G (2001) The mammalian circadian clock shop. *Seminars in Cell and Developmental Biology* 12:295–303.

6. Gizowski C, Zaelzer C, Bourque CW (2016) Clock-driven vasopressin neurotransmission mediates anticipatory thirst prior to sleep. *Nature* 537:685–688.

7. Ueta Y, Fujihara H, Serino R, Dayanithi G, Ozawa H, Matsuda K, Kawata M et al. (2005) Transgenic expression of enhanced green fluorescent protein enables direct visualization for physiological studies of vasopressin neurons and isolated nerve terminals of the rat. *Endocrinology* 146:406–413.

8. Tsuji T, Allchorne AJ, Zhang M, Tsuji C, Tobin VA, Pineda R, Raftogianni A et al. (2017) Vasopressin casts light on the suprachiasmatic nucleus. *Journal of Physiology* 595:3497–3514.

9. Tessmar-Raible K, Raible F, Christodoulou F, Guy K, Rembold M, Hausen H, Arendt D (2007) Conserved sensory-neurosecretory cell types in annelid and fish forebrain: insights into hypothalamus evolution. *Cell* 129:1389–1400.

10. Pfeffer M, Korf HW, Wicht H (2017) Synchronizing effects of melatonin on diurnal and circadian rhythms. *General and Comparative Endocrinology* pii: S0016-6480(17)30172-7. doi: 10.1016/j.ygcen.2017.05.013.

11. Foster RG, Kreitzman L (2014) The rhythms of life: what your body clock means to you! *Experimental Physiology* 99:599–606.

Chapter 16

1. Scott, Sir Walter (1833) *The Waverley Anecdotes*. Volume 2. Carter Hendee, 237.

2. Grigg D (1999) The changing geography of world food consumption in the second half of the twentieth century. *Geographical Journal* 165:1–11.

3. NCD Risk Factor Collaboration (NCD-RisC) (2016) Trends in adult body-mass index in 200 countries from 1975 to 2014: a pooled analysis of 1698 population-based measurement studies with 19.2 million participants. *Lancet* 387:1377–1396.

4. Ezzati M, Lopez AD, Rodgers A, Vander Hoorn S, Murray CJ; Comparative Risk Assessment Collaborating Group (2002) Selected major risk factors and global and regional burden of disease. *Lancet* 360:1347–1360.

5. Wright JD, Kennedy-Stephenson J, Wang CY, McDowell MA, Johnson CL (2004) Trends in intake of energy and macronutrients—United States, 1971–2000. *JAMA* 291:1193–1194.

6. Gregg EW, Cheng YJ, Cadwell BL, Imperatore G, Williams DE, Flegal KM, Narayan KM, Williamson DF (2005) Secular trends in cardiovascular disease risk factors according to Body Mass Index in US Adults. *JAMA* 293:1868–1874.

7. Heini AF, Weinsier RL (1997) Divergent trends in obesity and fat intake patterns: the American paradox. *American Journal of Medicine* 102:259–264.

8. Janssen I (2013) The public health burden of obesity in Canada. *Canadian Journal of Diabetes* 37:90–96.

9. Barclay A, Brand-Miller J (2011) The Australian paradox: a substantial decline in sugars intake over the same timeframe that overweight and obesity have increased. *Nutrients* 3: 491–504.

10. Annual statistics about food and drink purchases in the UK. Published March 9, 2017, by the Department for Environment, Food, and Rural Affairs. https://www.gov.uk/government/collections/family-food-statistics.

11. Prentice AM, Jebb SA (1995) Obesity in Britain: gluttony or sloth? *British Medical Journal* 311:437–439.

12. Wells HF, Busby JC (2008) Dietary assessment of major trends in U.S. food consumption, 1970–2005. *US Department of Agriculture Economic Information Bulletin Number 33.*

13. O'Rahilly S (2009) Human genetics illuminates the paths to metabolic disease. *Nature* 462:307–314.

14. Reddon H, Guéant JL, Meyre D (2016) The importance of gene-environment interactions in human obesity. *Clinical Science (London)* 130:1571–1597.

15. Hales CN, Barker DJ (2001) The thrifty phenotype hypothesis. *British Medical Bulletin* 60:5–20.

16. Speakman JR (2014) If body fatness is under physiological regulation, then how come we have an obesity epidemic? *Physiology (Bethesda)* 29:88–98.

17. Sellayah D, Cagampang FR, Cox RD (2014) On the evolutionary origins of obesity: a new hypothesis. *Endocrinology* 155:1573–1588.

18. Locke AE, Kahali B, Berndt SI et al. (2015) Genetic studies of body mass index yield new insights for obesity biology. *Nature* 518:197–206.

19. Ravelli G-P, Stein ZA, Susser MW (1976) Obesity in young men after famine exposure in utero and early infancy. *New England Journal of Medicine* 295:349–353.

20. Remacle C, Bieswal F, Bol V, Reusens B (2011) Developmental programming of adult obesity and cardiovascular disease in rodents by maternal nutrition imbalance. *American Journal of Clinical Nutrition* 94:1846S–1852S.

21. Orozco-Solís R, Matos RJB, Guzmán-Quevedo O, Lopes de Souza S, Bihouée A, Houlgatte R et al. (2010) Nutritional programming in the rat is linked to long-lasting changes in nutrient sensing and energy homeostasis in the hypothalamus. *PLoS One* 5: e13537.

22. Ng SF, Lin RC, Laybutt DR, Barres R, Owens JA, Morris MJ (2010) Chronic high-fat diet in fathers programs β-cell dysfunction in female rat offspring. *Nature* 467:963–966.

23. John GK, Mullin GE (2016) The gut microbiome and obesity. *Current Oncology Reports* 18:45.

24. Ridaura VK, Faith JJ, Rey FE, Cheng J, Duncan AE, Kau AL, Griffin NW et al. (2013) Gut microbiota from twins discordant for obesity modulate metabolism in mice. *Science* 341(6150):1241214. doi: 10.1126/science.1241214.

25. Cammarota G, Pecere S, Ianiro G, Masucci L, Currò D (2016) Principles of DNA-based gut microbiota assessment and therapeutic efficacy of fecal microbiota transplantation in gastrointestinal diseases. *Digestive Diseases* 34:279–285.

26. Marotz CA, Zarrinpar A (2016) Treating obesity and metabolic syndrome with fecal microbiota transplantation. *Yale Journal of Biology and Medicine* 89:383–388.

27. Villablanca PA, Alegria JR, Mookadam F, Holmes DR Jr, Wright RS, Levine JA (2015) Nonexercise activity thermogenesis in obesity management. *Mayo Clinic Proceedings* 90:509–519.

28. Johnson F, Mavrogianni A, Ucci M, Vidal-Puig A, Wardle J (2011) Could increased time spent in a thermal comfort zone contribute to population increases in obesity? *Obesity Reviews* 12:543–551.

29. Yang HK, Han K, Cho J-H, Yoon K-H, Cha B-Y, Lee S-H (2015) Ambient temperature and prevalence of obesity: a nationwide population-based study in Korea. *PLoS One* 10(11): e0141724.

30. Keith SW, Redden DT, Katzmarzyk PT, Boggiano MM, Hanlon EC, Benca RM et al. (2006) Putative contributors to the secular increase in obesity: exploring the roads less traveled. *International Journal of Obesity (London)* 30:1585–1594.

31. Church TS, Martin CK, Thompson AM, Earnest CP, Mikus CR, Blair SN (2009) Changes in weight, waist circumference and compensatory responses with different

doses of exercise among sedentary, overweight postmenopausal women. *PLoS One* 4(2): e4515.

32. Molé PA (1990) Impact of energy intake and exercise on resting metabolic rate. *Sports Medicine* 10:72–87.

33. Health Canada: Estimated Energy Requirements. https://www.canada.ca/en/health-canada/services/food-nutrition/canada-food-guide/food-guide-basics/estimated-energy-requirements.html. Accessed 03/11/2017.

34. Togo P, Osler M, Sørensen TI, Heitmann BL (2001) Food intake patterns and body mass index in observational studies. *International Journal of Obesity and Related Metabolic Disorders* 25:1741–1751.

Chapter 17

1. Khera R, Murad MH, Chandar AK, Dulai PS, Wang Z, Prokop LJ, Loomba R, Camilleri M, Singh S (2016) Association of pharmacological treatments for obesity with weight loss and adverse events: a systematic review and meta-analysis. *JAMA* 315:2424–2434.

2. Stafford RS, Radley DC (2003) National trends in antiobesity medication use. *Archives of Internal Medicine* 163:1046–1050.

3. Weintraub M (1992) Long-term weight control study: conclusions. *Clinical Pharmacology and Therapeutics* 51:642–646.

4. Abenhaim L, Moride Y, Brenot F, Rich S, Benichou J, Kurz X, Higenbottam T, Oakley C, Wouters E, Aubier M, Simonneau G, Bégaud B (1996) Appetite-suppressant drugs and the risk of primary pulmonary hypertension. International Primary Pulmonary Hypertension Study Group. *New England Journal of Medicine* 335:609–616.

5. Cunningham JW, Wiviott SD (2014) Modern obesity pharmacotherapy: weighing cardiovascular risk and benefit. *Clinical Cardiology* 37:693–699.

6. Narayanaswami V, Dwoskin LP (2017) Obesity: current and potential pharmacotherapeutics and targets. *Pharmacology and Therapeutics* 170:116–147.

7. Krentz AJ, Fujioka K, Hompesch M (2016) Evolution of pharmacological obesity treatments: focus on adverse side-effect profiles. *Diabetes, Obesity and Metabolism* 18:558–570.

8. Anderson EJ, Çakir I, Carrington SJ, Cone RD, Ghamari-Langroudi M, Gillyard T, Gimenez LE, Litt MJ. (2016) 60 years of POMC. Regulation of feeding and energy homeostasis by α-MSH. *Journal of Molecular Endocrinology* 56:T157–174.

9. Abdel-Malek ZA, Swope VB, Starner RJ, Koikov L, Cassidy P, Leachman S (2014) Melanocortins and the melanocortin 1 receptor, moving translationally toward melanoma prevention. *Archives of Biochemistry and Biophysics* 563:4–12.

10. Lane AM, McKay JT, Bonkovsky HL (2016) Advances in the management of erythropoietic protoporphyria—role of afamelanotide. *Application of Clinical Genetics* 9:179–189.

11. Hadley ME (2005) Discovery that a melanocortin regulates sexual functions in male and female humans. *Peptides* 26:1687–1689.

12. Caquineau C, Leng G, Douglas AJ (2012) Sexual behavior and neuronal activation in the vomeronasal pathway and hypothalamus of food-deprived male rats. *Journal of Neuroendocrinology* 24:712–723.

13. Flavell SW, Pokala N, Macosko EZ, Albrecht DR, Larsch J, Bargmann CI (2013) Serotonin and the neuropeptide PDF initiate and extend opposing behavioral states in *C. elegans*. *Cell* 154:1023–1035.

14. Burke LK, Heisler LK (2015) 5-hydroxytryptamine medications for the treatment of obesity. *Journal of Neuroendocrinology* 27:389–398.

15. Vu JP, Larauche M, Flores M, Luong L, Norris J, Oh S, Liang LJ, Waschek J, Pisegna JR, Germano PM (2015) Regulation of appetite, body composition, and metabolic hormones by vasoactive intestinal polypeptide (VIP). *Journal of Molecular Neuroscience* 56:377–387.

16. Bray GA, Frühbeck G, Ryan DH, Wilding JP (2016) Management of obesity. *Lancet* 387:1947–1956.

17. Pocai A (2013) Action and therapeutic potential of oxyntomodulin. *Molecular Metabolism* 3:241–51.

18. Prinz P, Stengel A (2017) Control of food intake by gastrointestinal peptides: mechanisms of action and possible modulation in the treatment of obesity. *Journal of Neurogastroenterology and Motility* 23:180–196.

19. Tan T, Behary P, Tharakan G, Minnion J, Al-Najim W, Wewer Albrechtsen NJ, Holst JJ, Bloom SR (2017) The effect of a subcutaneous infusion of GLP-1, OXM and PYY on energy intake and expenditure in obese volunteers. *Journal of Clinical Endocrinology and Metabolism*. doi: 10.1210/jc.2017-00469.

Chapter 18

1. Speakman JR, Levitsky DA, Allison DB, Bray MS, de Castro JM, Clegg DJ, Clapham JC et al. (2011) Set points, settling points and some alternative models: theoretical options to understand how genes and environments combine to regulate body adiposity. *Disease Models and Mechanisms* 4:733–745.

2. Hervey GR (1969) Regulation of energy balance. *Nature* 222:629–631.

3. Han PW, Mu JY, Lepkovsky S (1963) Food intake of parabiotic rats. *American Journal of Physiology* 205:1139–1143.

4. Farooqi IS, O'Rahilly S (2000) Recent advances in the genetics of severe childhood obesity. *Archives of Disease in Childhood* 83:31–34.

5. Coleman DL, Hummel KP (1969) Effects of parabiosis of normal with genetically diabetic mice. *American Journal of Physiology* 217:1298–1304.

6. Coleman DL (1973) Effects of parabiosis of obese with diabetes and normal mice. *Diabetologia* 9:294–298.

7. Harris RB, Hervey E, Hervey GR, Tobin G (1987) Body composition of lean and obese Zucker rats in parabiosis. *International Journal of Obesity* 11:275–283.

8. Harris RB (2013) Is leptin the parabiotic "satiety" factor? Past and present interpretations. *Appetite* 61:111–118.

9. Coleman DL (2010) A historical perspective on leptin. *Nature Medicine* 16:1097–1099.

10. Kissileff HR (1991) Chance and necessity in ingestive behavior. *Appetite* 17:1–22.

11. Zhang, Y, Proenca R, Maffei M, Barone M, Leopold L, Friedman JM (1994) Positional cloning of the mouse obese gene and its human homolog. *Nature* 372:425–432.

12. Phillips MS, Liu Q, Hammond HA, Dugan V, Hey PJ, Caskey CJ, Hess JF (1996) Leptin receptor missense mutation in the fatty Zucker rat. *Nature Genetics* 13:18–19.

13. Kojima M, Hosoda H, Date Y, Nakazato M, Matsuo H, Kangawa K (1999) Ghrelin is a growth-hormone-releasing acylated peptide from stomach. *Nature* 402:656–660.

14. Perello M, Dickson SL (2015) Ghrelin signaling on food reward: a salient link between the gut and the mesolimbic system. *Journal of Neuroendocrinology* 27:424–434.

15. Leng G, Adan RA, Belot M, Brunstrom JM, de Graaf K, Dickson SL, Hare T et al. (2016) The determinants of food choice. *Proceedings of the Nutrition Society* 1:1–12.

16. Heiman ML, Ahima RS, Craft LS, Schoner B, Stephens TW, Flier JS. (1997) Leptin inhibition of the hypothalamic-pituitary-adrenal axis in response to stress. *Endocrinology* 138:3859–3863.

17. Sonneville KR, Horton NJ, Micali N, Crosby RD, Swanson SA, Solmi F, Field AE. (2013) Longitudinal associations between binge eating and overeating and adverse outcomes among adolescents and young adults: does loss of control matter? *Journal of the American Medical Association Pediatrics* 167:149–155.

18. Neumark-Sztainer D, Wall M, Story M, Standish AR (2012) Dieting and unhealthy weight control behaviors during adolescence: associations with 10-year changes in body mass index. *Journal of Adolescent Health* 50:80–86.

19. Badman MK, Flier JS (2007) The adipocyte as an active participant in energy balance and metabolism. *Gastroenterology* 132:2103–2115.

20. Coll AP, Farooqi IS, O'Rahilly S (2007) The hormonal control of food intake. *Cell* 129:251–262.

21. Trayhurn P, Beattie JH. (2001) Physiological role of adipose tissue: white adipose tissue as an endocrine and secretory organ. *Proceedings of the Nutrition Society* 60:329–339.

22. Hazell TJ, Islam H, Townsend LK, Schmale MS, Copeland JL (2016) Effects of exercise intensity on plasma concentrations of appetite-regulating hormones: potential mechanisms. *Appetite* 98:80–88.

23. Mosialou I, Shikhel S, Liu JM, Maurizi A, Luo N, He Z, Huang Y et al. (2017) MC4R-dependent suppression of appetite by bone-derived lipocalin 2. *Nature* 543:385–390.

24. Shipp SL, Cline MA, Gilbert ER (2016) Recent advances in the understanding of how neuropeptide Y and α-melanocyte stimulating hormone function in adipose physiology. *Adipocyte* 5:333–350.

Chapter 19

1. This chapter is based on primary papers reviewed in Leng G, Sabatier N (2017) Oxytocin—the sweet hormone? *Trends in Endocrinology and Metabolism* 28:365–376.

2. Johnstone LE, Fong TM, Leng G (2006) Neuronal activation in the hypothalamus and brainstem during feeding in rats. *Cell Metabolism* 4:313–321.

3. Tung YC, Ma M, Piper S, Coll A, O'Rahilly S, Yeo GS (2008) Novel leptin-regulated genes revealed by transcriptional profiling of the hypothalamic paraventricular nucleus. *Journal of Neuroscience* 28:12419–12426.

4. Billig I, Yates BJ, Rinaman L (2001) Plasma hormone levels and central c-Fos expression in ferrets after systemic administration of cholecystokinin. *American Journal of Physiology* 281:R1243–1255.

5. Lincoln DW, Renfree MB (1981) Milk ejection in a marsupial, *Macropus agilis*. *Nature* 289:504–506.

6. Verbalis JG, Mangione MP, Stricker EM (1991) Oxytocin produces natriuresis in rats at physiological plasma concentrations. *Endocrinology* 128:1317–1322.

Chapter 20

1. Starling EH (1923) The wisdom of the body. *British Medical Journal* 2:685–690.

2. Agnati LF, Zoli M, Strömberg I, Fuxe K (1995) Intercellular communication in the brain: wiring versus volume transmission. *Neuroscience* 69:711–726.

3. Bolam JP, Pissadaki EK (2012) Living on the edge with too many mouths to feed: why dopamine neurons die. *Movement Disorders* 27:1478–1483.

4. Moss J, Bolam JP (2008) A dopaminergic axon lattice in the striatum and its relationship with cortical and thalamic terminals. *Journal of Neuroscience* 28:11221–11230.

5. Bakker J, De Mees C, Douhard Q et al. (2006) α-Fetoprotein protects the developing female mouse brain from masculinization and defeminization by estrogens. *Nature Neuroscience* 9:220–226.

6. Raisman G, Field PM (1973) Sexual dimorphism in the neuropil of the preoptic area of the rat and its dependence on neonatal androgen. *Brain Research* 54:1–29.

7. Hofman MA, Swaab DF (1993) Diurnal and seasonal rhythms of neuronal activity in the suprachiasmatic nucleus of humans. *Journal of Biological Rhythms* 8:283–295.

8. Swaab DF, Hofman MA (1990) An enlarged suprachiasmatic nucleus in homosexual men. *Brain Research* 537:141–148.

9. Swaab DF, Slob AK, Houtsmuller EJ, Brand T, Zhou JN (1995) Increased number of vasopressin neurons in the suprachiasmatic nucleus (SCN) of "bisexual" adult male rats following perinatal treatment with the aromatase blocker ATD. *Developmental Brain Research* 85:273–279.

10. Quoted in Henriksen JH, Schaffalitzky de Muckadell OB (2000) Secretin, its discovery, and the introduction of the hormone concept. *Scandinavian Journal of Clinical Laboratory Investigation* 60:463–472.

11. Branco T, Staras K (2009) The probability of neurotransmitter release: variability and feedback control at single synapses. *Nature Reviews Neuroscience* 10:373–383.

12. Yuan T, Lu J, Zhang J, Zhang Y, Chen L (2015) Spatiotemporal detection and analysis of exocytosis reveal fusion "hotspots" organized by the cytoskeleton in endocrine cells. *Biophysical Journal* 108:251–260.

Chapter 21

1. Brontë, Emily (1847/1994) *Wuthering Heights*. Penguin Popular Classics, 73.

2. Leng G, Meddle SL, Douglas AJ (2008) Oxytocin and the maternal brain. *Current Opinion in Pharmacology* 8:731–734.

3. Pedersen CA, Prange AJ Jr (1979) Induction of maternal behavior in virgin rats after intracerebroventricular administration of oxytocin. *Proceedings of the National Academy of Sciences USA* 76:6661–6665.

4. Marlin BJ, Mitre M, D'amour JA, Chao MV, Froemke RC (2015) Oxytocin enables maternal behavior by balancing cortical inhibition. *Nature* 520:499–504.

5. Da Costa AP, Guevara-Guzman RG, Ohkura S, Goode JA, Kendrick KM (1996) The role of oxytocin release in the paraventricular nucleus in the control of maternal behavior in the sheep. *Journal of Neuroendocrinology* 8:163–177.

6. Lévy F, Kendrick KM, Keverne EB, Piketty V, Poindron P (1992) Intracerebral oxytocin is important for the onset of maternal behavior in inexperienced ewes delivered under peridural anesthesia. *Behavioral Neuroscience* 106:427–432.

7. Knobloch HS, Charlet A, Hoffmann LC, Eliava M, Khrulev S, Cetin AH, Osten P et al. (2012) Evoked axonal oxytocin release in the central amygdala attenuates fear response. *Neuron* 73:553–566.

8. Insel TR (1992) Oxytocin—a neuropeptide for affiliation: evidence from behavioral, receptor autoradiographic, and comparative studies. *Psychoneuroendocrinology* 17:3–35.

9. Williams JR, Insel TR, Harbaugh CR, Carter CS (1994) Oxytocin administered centrally facilitates formation of a partner preference in female prairie voles (*Microtus ochrogaster*). *Journal of Neuroendocrinology* 6:247–250.

10. Harris GW, Pickles VR (1953) Reflex stimulation of the neurohypophysis (posterior pituitary gland) and the nature of posterior pituitary hormone(s). *Nature* 172:1049.

11. Insel TR, Preston S, Winslow JT (1995) Mating in the monogamous male: behavioral consequences. *Physiology and Behavior* 57:615–627.

12. Tobin VA, Hashimoto H, Wacker DW, Takayanagi Y, Langnaese K, Caquineau C, Noack J et al. (2010) An intrinsic vasopressin system in the olfactory bulb is involved in social recognition. *Nature* 464:413–417.

13. Young LJ, Wang Z, Cooper TT, Albers H (2000) Vasopressin (V1a) receptor binding, mRNA expression and transcriptional regulation by androgen in the Syrian hamster brain. *Journal of Neuroendocrinology* 12:1179–1185.

14. Young LJ, Lim MM, Gingrich B, Insel TR (2001) Cellular mechanisms of social attachment. *Hormones and Behavior* 40:133–138.

15. Pitkow LJ, Sharer CA, Ren X, Insel TR, Terwilliger EF, Young LJ (2001) Facilitation of affiliation and pair-bond formation by vasopressin receptor gene transfer into the ventral forebrain of a monogamous vole. *Journal of Neuroscience* 21:7392–7396.

16. Young LJ, Nilsen R, Waymire KG, MacGregor GR, Insel TR (1999) Increased affiliative response to vasopressin in mice expressing the V1a receptor from a monogamous vole. *Nature* 400:766–768.

17. Johnson ZV, Young LJ (2017) Oxytocin and vasopressin neural networks: implications for social behavioral diversity and translational neuroscience. *Neuroscience and Biobehavioral Reviews* 76:87–98.

18. Herkenham M (1987) Mismatches between neurotransmitter and receptor localizations in brain: observations and implications. *Neuroscience* 23:1–38.

19. Harte SE, Meyers JB, Donahue RR, Taylor BK, Morrow TJ (2016) Mechanical Conflict System: a novel operant method for the assessment of nociceptive behavior. *PLoS One* 11(2):e0150164.

20. Merullo DP, Cordes MA, DeVries SM, Stevenson SA, Riters LV (2015) Neurotensin neural mRNA expression correlates with vocal communication and other highly-motivated social behaviors in male European starlings. *Physiology and Behavior* 151:155–161.

21. Fadok JP, Krabbe S, Markovic M, Courtin J, Xu C, Massi L, Botta P et al. (2017) A competitive inhibitory circuit for selection of active and passive fear responses. *Nature* 542:96–100.

22. Puga L, Alcántara-Alonso V, Coffeen U, Jaimes O, de Gortari P (2016) TRH injected into the nucleus accumbens shell releases dopamine and reduces feeding motivation in rats. *Behavioral Brain Research* 306:128–136.

23. Wang L, Hiller H, Smith JA, de Kloet AD, Krause EG. (2016) Angiotensin type 1a receptors in the paraventricular nucleus of the hypothalamus control cardiovascular reactivity and anxiety-like behavior in male mice. *Physiological Genomics* 48:667–760.

24. Han S, Soleiman MT, Soden ME, Zweifel LS, Palmiter RD (2015) Elucidating an affective pain circuit that creates a threat memory. *Cell* 162:363–374.

25. Wilson J, Markie D, Fitches A (2012) Cholecystokinin system genes: associations with panic and other psychiatric disorders. *Journal of Affective Disorders* 136:902–908.

26. Frieling H, Bleich S, Otten J, Römer KD, Kornhuber J, de Zwaan M, Jacoby GE, Wilhelm J, Hillemacher T (2008) Epigenetic downregulation of atrial natriuretic peptide but not vasopressin mRNA expression in females with eating disorders is related to impulsivity. *Neuropsychopharmacology* 33:2605–2609.

27. Jorde A, Bach P, Witt SH, Becker K, Reinhard I, Vollstädt-Klein S, Kirsch M et al. (2014) Genetic variation in the atrial natriuretic peptide transcription factor GATA4 modulates amygdala responsiveness in alcohol dependence. *Biological Psychiatry* 75:790–797.

28. Nummenmaa L, Tuominen L (2017) Opioid system and human emotions. *British Journal of Pharmacology*. doi: 10.1111/bph.13812.

29. Young LJ, Winslow JT, Nilsen R, Insel TR (1997) Species differences in V1a receptor gene expression in monogamous and nonmonogamous voles: behavioral consequences. *Behavioral Neuroscience* 111:599–605.

30. Ludwig M, Leng G (2006) Dendritic peptide release and peptide-dependent behaviors. *Nature Reviews Neuroscience* 7:126–36.

31. Geerling JC, Loewy AD (2008) Central regulation of sodium appetite. *Experimental Physiology* 93:177–209.

32. Smith CM, Walker LL, Chua BE, McKinley MJ, Gundlach AL, Denton DA, Lawrence AJ (2015) Involvement of central relaxin-3 signaling in sodium (salt) appetite. *Experimental Physiology* 100:1064–1072.

33. Alvares GA, Quintana DS, Whitehouse AJ (2017) Beyond the hype and hope: Critical considerations for intranasal oxytocin research in autism spectrum disorder. *Autism Research* 10:25–41.

34. Leng G, Ludwig M (2016) Intranasal oxytocin: myths and delusions. *Biological Psychiatry* 79:243–250.

35. Nave G, Camerer C, McCullough M (2015) Does oxytocin increase trust in humans? A critical review of research. *Perspectives in Psychological Sciences* 10:772–789.

36. Walum H, Waldman ID, Young LJ (2016) Statistical and methodological considerations for the interpretation of intranasal oxytocin studies. *Biological Psychiatry* 79:251–257.

37. Ioannidis JP (2005) Why most published research findings are false. *PLoS Medicine* 2(8):e124.

Chapter 22

1. Smith, Adam (1795) The principles which lead and direct philosophical enquiries; illustrated by the history of the ancient physics. In *Essays on Philosophical Subjects*. Dublin, 134

2. Henry J, Cornet V, Bernay B, Zatylny-Gaudin C (2013) Identification and expression of two oxytocin/vasopressin-related peptides in the cuttlefish *Sepia officinalis*. *Peptides* 46:159–166.

3. Braida D, Donzelli A, Martucci R, Capurro V, Busnelli M, Chini B, Sala M (2012) Neurohypophyseal hormones manipulation modulate social and anxiety-related behavior in zebrafish. *Psychopharmacology (Berlin)* 220:319–330.

4. Palazzo AF, Gregory TR (2014) The case for junk DNA. *PLoS Genetics* 10(5):e1004351.

5. Paré P, Paixão-Côrtes VR, Tovo-Rodrigues L, Vargas-Pinilla P, Viscardi LH, Salzano FM, Henkes LE, Bortolini MC (2016) Oxytocin and arginine vasopressin receptor evolution: implications for adaptive novelties in placental mammals. *Genetics and Molecular Biology* 39:646–657.

6. Mayasich SA, Clarke BL (2016) The emergence of the vasopressin and oxytocin hormone receptor gene family lineage: clues from the characterization of vasotocin

receptors in the sea lamprey (*Petromyzon marinus*). *General and Comparative Endocrinology* 226:88–101.

7. Kazazian HH Jr, Moran JV (1998) The impact of L1 retrotransposons on the human genome. *Nature Genetics* 19:19–24.

8. Murphy D, Si-Hoe SL, Brenner S, Venkatesh B (1998) Something fishy in the rat brain: molecular genetics of the hypothalamo-neurohypophysial system. *Bioessays* 20:741–749.

9. Eaton JL, Holmqvist B, Glasgow E (2008) Ontogeny of vasotocin-expressing cells in zebrafish: selective requirement for the transcriptional regulators *orthopedia* and *single-minded 1* in the preoptic area. *Developmental Dynamics* 237:995–1005.

10. Larson ET, O'Malley DM, Melloni RH Jr (2006) Aggression and vasotocin are associated with dominant-subordinate relationships in zebrafish. *Behavioral Brain Research* 167: 94–102.

11. Gainer H (2012) Cell-type specific expression of oxytocin and vasopressin genes: an experimental odyssey. *Journal of Neuroendocrinology* 24:528–538.

12. Lam BYH, Cimino I, Polex-Wolf J, Nicole Kohnke S, Rimmington D, Iyemere V, Heeley N et al. (2017) Heterogeneity of hypothalamic pro-opiomelanocortin-expressing neurons revealed by single-cell RNA sequencing. *Molecular Metabolism* 6:383–392.

13. Zaelzer C, Hua P, Prager-Khoutorsky M, Ciura S, Voisin DL, Liedtke W, Bourque CW (2015) ΔN-TRPV1: A molecular co-detector of body temperature and osmotic stress. *Cell Reports* 13:23–30.

14. Hutton J (1788) Theory of the earth. *Transactions of the Royal Society of Edinburgh* 1:304.

Chapter 23

1. Hutton, James (1794) *An Investigation of the Principles of Knowledge.* Volume 2. A. Strahan and T. Cadell. The geologist James Hutton paved the way for Darwin's theory of evolution by natural selection. Hutton's own mind, opened by his recognition of deep time, came close to that theory. How close may be judged from this quotation, expressed though it is in Hutton's palsied prose.

2. Edelman GM, Gally JA (2001) Degeneracy and complexity in biological systems. *Proceedings of the National Academy of Sciences USA* 98:13763–13768.

3. Russell JA, Leng G, Douglas AJ (2003) The magnocellular oxytocin system, the fount of maternity: adaptations in pregnancy. *Frontiers in Neuroendocrinology* 24:27–61.

4. Gropp E, Shanabrough M, Borok E, Xu AW, Janoschek R, Buch T, Plum L et al. (2005) Agouti-related peptide-expressing neurons are mandatory for feeding. *Nature Neuroscience* 8:1289–1291.

5. Luquet S, Perez FA, Hnasko TS, Palmiter RD (2005) NPY/AgRP neurons are essential for feeding in adult mice but can be ablated in neonates. *Science* 310:683–685.

6. Wu Q, Boyle MP, Palmiter RD (2009) Loss of GABAergic signaling by AgRP neurons to the parabrachial nucleus leads to starvation. *Cell* 137:1225–1234.

7. Wortley KE, Anderson KD, Garcia K, Murray JD, Malinova L, Liu R, Moncrieffe M et al. (2004) Genetic deletion of ghrelin does not decrease food intake but influences metabolic fuel preference. *Proceedings of the National Academy of Sciences USA* 101:8227–8232.

8. Roizen J, Luedke CE, Herzog ED, Muglia LJ (2007) Oxytocin in the circadian timing of birth. *PLoS One.* 2(9):e922. Oxytocin-deficient mice normally give birth at about the same time as normal mice, but this is disrupted when there is a disturbance in the light-dark cycle.

9. Russell JA, Leng G (1998) Sex, parturition and motherhood without oxytocin? *Journal of Endocrinology* 157:343–359.

10. Yu F, Jiang QJ, Sun XY, Zhang RW (2015) A new case of complete primary cerebellar agenesis: clinical and imaging findings in a living patient. *Brain* 138(Pt 6):e353.

11. Feuillet L, Dufour H, Pelletier J (2007) Brain of a white-collar worker. *Lancet* 370:262.

Chapter 24

1. Scott, Walter (1808) *Marmion; A Tale of Flodden Field.* Project Gutenberg http://www.gutenberg.org/ebooks/5077.

2. *"I know that the right kind of leader for the Labour Party is a kind of desiccated calculating-machine who must not in any way permit himself to be swayed by indignation."* Spoken in 1954 by the Welsh Labour Party politician Aneurin Bevan (1897–1960). *Brewer's Dictionary of Modern Phrase & Fable* (2 ed.) Edited by John Ayto and Ian Crofton 2009. Chambers Harrap.

3. FitzHugh R (1960) Thresholds and plateaus in the Hodgkin-Huxley nerve equations. *Journal of General Physiology* 43:867–896.

4. FitzHugh R (1966) An electronic model of the nerve membrane for demonstration purposes. *Journal of Applied Physiology* 21:305–308.

Index

Printed in the United States
by Baker & Taylor Publisher Services